Sparsity Methods for Systems and Control

Masaaki Nagahara
The University of Kitakyushu, Japan

now

the essence of knowledge

Published, sold and distributed by:
now Publishers Inc.
PO Box 1024
Hanover, MA 02339
United States
Tel. +1-781-985-4510
www.nowpublishers.com
sales@nowpublishers.com

Outside North America:
now Publishers Inc.
PO Box 179
2600 AD Delft
The Netherlands
Tel. +31-6-51115274

ISBN: 978-1-68083-724-7
E-ISBN: 978-1-68083-725-4
DOI: 10.1561/9781680837254

Table of Contents

Preface

Scientists and engineers love simplicity. We prefer simple laws of nature like Newton's laws of motion to a complicated law with tens of equations, simple word and phrases to explain the nature of life to books with hundreds of pages, and a simple room just with a vessel with flowers to a colorful room full of furniture. Simple is best.

But simple is not easy. It is rather very difficult. Simplification is probably the most difficult problem of design. In this book, we will see how to solve this type of problems in engineering. The idea is to introduce the sparsity. It makes the hard problem solvable.

The method of sparsity becomes more and more popular in engineering, in particular, in signal processing, machine learning, statistics, etc. It is known as compressed sensing, compressive sampling, sparse representation, or sparse modeling. More recently, this method has been applied to systems and control to design resource-aware control systems. This book gives a comprehensive guide to sparsity methods for systems and control, from standard sparsity methods in finite-dimensional vector spaces (Part I) to optimal control methods in infinite-dimensional function spaces (Part II).

The primary objective of this book is to give *how* to use sparsity methods for several engineering problems. I omitted theoretical aspects of the sparsity methods, such as characterization of sparse solutions based on random matrix theory and convergence of optimization algorithms. If you are interested in those theoretical aspects, please refer to the references listed in the section titled "Further reading" at the end of each chapter. Instead, I provide MATLAB programs by which you can try sparsity methods by yourself. You will obtain deep understanding of sparsity methods by running these MATLAB programs. MATLAB programs and other information on this book are found at

https://nagahara-masaaki.github.io/spm_en

The book is an English translation from the author's book *Sparse Modeling* published in 2017 by Corona Publishing, Japan. Also, the contents of this book have been updated and added based on two university courses: *Sparsity Methods for Systems and Control* (SC637) taught at Indian Institute of Technology (IIT) Bombay in 2018 during my sabbatical, and *Sparse Modeling* (M153F31) at The University of Kitakyushu, Japan during 2018–2020. These courses are for graduate students, but I believe undergraduate students can read with basic knowledge of linear algebra and elementary calculus. Also, this book (especially Part II) appeals to professional researchers and engineers who are interested in applying sparsity methods to systems and control. The courses start from finite-dimensional optimization (i.e. Part I) to optimal control (Part II), but if you want to quickly know about the optimal control, you can omit Part I and directly start from Chapter 7 (Part II). Chapter 1 is an introductory chapter, where I mainly mention the history of sparsity methods in engineering. You can omit (or read later) Chapter 1 since this chapter is completely independent of the other chapters.

Acknowledgment

I am grateful to Prof. Debasish Chatterjee for giving me the opportunity to teach a course on this topic in IIT Bombay. This triggered me to write this book. I would like to thank students who took my courses in IIT Bombay for finding a lot of typos and mistakes not only in English grammar but also in equations and mathematical proofs.

Masaaki Nagahara
Kitakyushu, Japan

Notation

A finite-dimensional vector is represented in a bold face, e.g. $\boldsymbol{x}$, when the size of the vector is greater than 2. For one-dimensional vectors, we do not use a bold face and simply write like x, regarding as a scalar.

We denote by $\mathbb{R}^n$ the set of n-dimensional real column vectors, and by $\mathbb{R}^{m \times n}$ the set of $m \times n$ real matrices. The transpose of a vector $\boldsymbol{x}$ and a matrix A are respectively denoted by $\boldsymbol{x}^\top$ or $A^\top$. The i-th element of a vector $\boldsymbol{x}$ and the (i, j)-th element of a matrix A are respectively denoted by $(\boldsymbol{x})_i$ and $[A]_{ij}$. We denote by $\mathbb{Z}$ the set of integers and by $\mathbb{N}$ the set of natural numbers, that is, $\mathbb{N} = \{1, 2, 3, \ldots\}$.

For a vector $\boldsymbol{x} \in \mathbb{R}^n$, $\mathrm{supp}(\boldsymbol{x})$ denotes the support set of $\boldsymbol{x}$, that is, the set of nonzero elements of $\boldsymbol{x} = [x_1, \ldots, x_n]^\top \in \mathbb{R}^n$:

$$\mathrm{supp}(\boldsymbol{x}) \triangleq \{i \in \{1, \ldots, n\} : x_i \neq 0\}. \tag{1}$$

The ℓ^0 norm of $\boldsymbol{x} \in \mathbb{R}^n$ is defined by

$$\|\boldsymbol{x}\|_0 \triangleq \#\big(\mathrm{supp}(\boldsymbol{x})\big), \tag{2}$$

where $\#(\cdot)$ returns the number of elements of the argument set. The ℓ^p norm with $p \geq 1$ is defined by

$$\|\boldsymbol{x}\|_p \triangleq \left\{\sum_{i=1}^{n} |x_i|^p\right\}^{1/p}, \tag{3}$$

and the ℓ^∞ norm by

$$\|\boldsymbol{x}\|_\infty \triangleq \max_{i=1,2,\ldots,n} |x_i|. \tag{4}$$

In Part II, these norms will be denoted by $\|\boldsymbol{x}\|_{\ell^0}$, $\|\boldsymbol{x}\|_{\ell^p}$, and $\|\boldsymbol{x}\|_{\ell^\infty}$ to distinguish norms for continuous-time signals.

For a vector $x \in \mathbb{R}^n$, and an index set $S \subset \{1, 2, \ldots, n\}$, we denote by x_S the restriction of x to S. If $x = [x_1, x_2, \ldots, x_n]^\top$ and $S = \{i_1, i_2, \ldots, i_k\}$ ($1 \le i_1 < i_2 < \cdots < i_k \le n$), then

$$x_S = [x_{i_1}, x_{i_2}, \ldots, x_{i_k}]^\top \in \mathbb{R}^k. \tag{5}$$

Also, for $\Phi = [\phi_1, \phi_2, \ldots, \phi_n] \in \mathbb{R}^{m \times n}$, Φ_S is defined as

$$\Phi_S = [\phi_{i_1}, \phi_{i_2}, \ldots, \phi_{i_k}] \in \mathbb{R}^{m \times k}. \tag{6}$$

The complement of a set S is denoted by S^c.

Let $f : [0, T] \to \mathbb{R}$ be a measurable function with $T > 0$. The support of f is denoted by $\mathrm{supp}(f)$ and defined by

$$\mathrm{supp}(f) \triangleq \{t \in [0, T] : f(t) \ne 0\}. \tag{7}$$

The L^0 norm of f is defined by

$$\|f\|_0 \triangleq \mu(\mathrm{supp}(f)), \tag{8}$$

where μ is the Lebesgue measure over $\mathbb{R}$. The L^p norm with $p \ge 1$ is defined by

$$\|f\|_p \triangleq \left\{ \int_0^T |f(t)|^p dt \right\}^{1/p}, \tag{9}$$

and the L^∞ norm by

$$\|f\|_\infty \triangleq \sup_{t \in [0,T]} |f(t)|. \tag{10}$$

We denote by $L^p(0, T)$ with $p \ge 1$ or $p = \infty$ the set of functions with finite L^p norm.

For a function $f : \mathbb{R}^n \to \mathbb{R}$, the *gradient* ∇f is defined by

$$\nabla f \triangleq \frac{\partial f}{\partial x} = \left[\frac{\partial f}{x_1}, \frac{\partial f}{x_2}, \ldots, \frac{\partial f}{x_n} \right]^\top \in \mathbb{R}^n. \tag{11}$$

We say a real-valued function $f(n)$, $n \in \mathbb{N}$, is $O(g(n))$ if

$$\limsup_{n \to \infty} \left| \frac{f(n)}{g(n)} \right| < \infty.$$

DOI: 10.1561/9781680837254.ch1

Chapter 1

Introduction

In this chapter, we briefly review the history of sparsity methods in science and engineering. The chapter will motivate you to learn this topic. The content of this chapter is independent of the other chapters, and readers interested in the technical aspects of sparsity methods can skip this chapter without much effect on their understanding of the rest of the book.

1.1 Occam's Razor

At the root of sparsity methods is the idea that one should not assume more than is necessary to explain certain things. This is known as *Occam's razor*, also called the *law of parsimony*, developed by Ockham in the 14th century. This idea was not invented by Ockham, but rather long before him, for example, by Claudius Ptolemy (90AD–168AD) and Aristotle (384BC–322BC). This is a very familiar concept to us, especially in Japan, where there is a culture of Zen and Wabi/Sabi, which can be roughly translated as *simple is best*.

There is a satirical depiction of the opposite of Occam's razor, *Rube Goldberg's machine*. Figure 1.1 shows an example of Goldberg's machine. The machine is

1

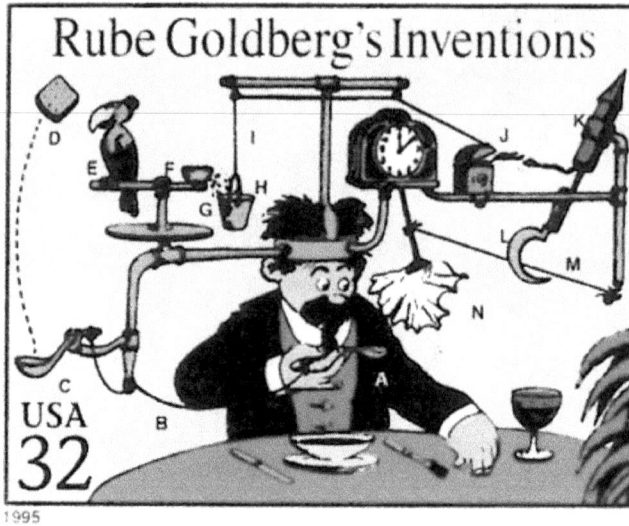

Figure 1.1. Goldberg's machine (self-operating napkin).

"self-operating napkin," which automatically wipes off the dirt from the beard when he drinks soup. This caricature depicts a machine that performs very simple actions with extreme complexity, and satirizes the large-scale mechanization in the first half of the 20th century. The machine runs as follows.

1. The man raises the spoon of soup (A) to his mouth.
2. The string (B) attached to the spoon (A) is pulled.
3. The ladle (C) moves.
4. Cracker (D) flies on the parrot (E).
5. The parrot (E) takes off after the cracker (D).
6. The perch (F) tilts.
7. The seeds (G) on the perch (F) spills out, and goes into the pail (H).
8. The string (I) is pulled by the extra weight in the pail.
9. It ignites the cigar lighter (J).
10. The fuse of the rocket (K) is lit and it takes off.
11. The knife (L) attached to the rocket (K) cuts the string (M).
12. The pendulum swings and the napkin (N) wipes the dirt from the beard.

The Goldberg machine is obviously strange. However, in this highly technologized society, we might have created something like the Goldberg machine without even realizing it. The sparsity method is therefore an essential technique to avoid such a situation.

1.2 Group Testing

Group testing is one of the first attempts to apply a sparsity method to a scientific problem. Group testing was proposed by Robert Dorfman in 1948 as a problem of finding an infected person among a large number of patients in a small number of blood tests [32]. For example, suppose that only one of eight patients is infected with a disease, which can be detected by examining the blood. Now, we have eight blood samples from the eight patients. Since blood testing is expensive and time-consuming, we want to identify the infected person as few times as possible. In this case, there is a good way to do this (see also Figure 1.2).

- (TEST 1) We first divide the blood of the eight patients into two groups of four patients, and take a little bit of blood from each of the eight patients, and mix it for each group. Since there is only one infected person, the blood from either group will test positive.
- (TEST 2) Divide the group that tested positive into two groups of two patients, and do the same thing. At this point, the number of suspicious person has been narrowed down to two.
- (TEST 3) Finally, by examining the blood of the two individuals separately, the infected person can be uniquely identified.

By this method, it is possible to identify an infected person in six tests, whereas eight tests would be required for an individual blood test. In general, according to the above method, if there is only one infected person among 2^T people, we can

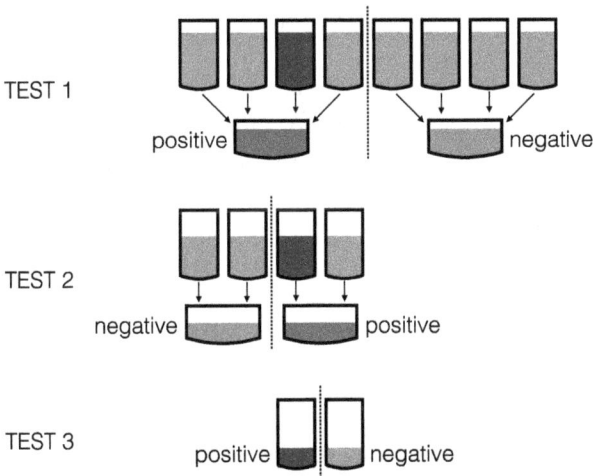

Figure 1.2. Group testing from eight blood samples.

identify the infected person in less than $2T$ tests. For example, for 1,024 patients, only 20 tests are needed to identify the infected person. We can see that group testing can dramatically reduce the number of tests compared to testing all patients' blood individually. We would like to consider a sophisticated method like this in a general situation where a few people in 100,000 for example are infected, instead of examining the blood of 100,000 people individually. This is the problem of group testing.

Now let us describe the problem of group testing in detail. Let n be the number of people to be tested. Define a variable x_i representing whether the i-th person ($i \in \{1, 2, \ldots, n\}$) is infected or not as

$$x_i \triangleq \begin{cases} 1, & \text{if the } i\text{-th person is infected,} \\ 0, & \text{otherwise.} \end{cases} \tag{1.1}$$

Define an n-dimensional binary vector that takes values of 0 or 1 as

$$\boldsymbol{x} \triangleq [x_1, x_2, \ldots, x_n]^\top \in \{0, 1\}^n, \tag{1.2}$$

where, $\{0, 1\}^n$ is the set of n-dimensional vectors whose elements are 0 or 1. The problem here is to find this n-dimensional binary vector. Of course, if we examine each one of them individually, we can determine the vector $\boldsymbol{x}$ with n tests, but here we want to identify $\boldsymbol{x}$ with a much smaller number of tests.

Let us consider a subset S of the set $\{1, 2, \ldots, n\}$. We define a function A that returns 1 if there is an infected person in S and 0 otherwise as follows:

$$A(S) = \begin{cases} 1, & \text{if } \sum_{i \in S} x_i \geq 1 \\ 0, & \text{otherwise.} \end{cases} \tag{1.3}$$

In group testing, a number of people are selected among n people to form a group, and their blood is mixed to test for infection. We form m groups denoted by $S_1, S_2, \ldots, S_m$. Define vector $\boldsymbol{y}$ of the results of the tests for these groups, $A(S_1), A(S_2), \ldots, A(S_m)$, by

$$\boldsymbol{y} \triangleq \left[A(S_1), A(S_2), \ldots, A(S_m) \right]^\top \in \{0, 1\}^m. \tag{1.4}$$

Also, we define matrix $\Phi \in \{0, 1\}^{m \times n}$ by

$$\Phi_{ij} \triangleq \begin{cases} 1, & \text{if } i \in S_j \\ 0, & \text{otherwise,} \end{cases} \tag{1.5}$$

where Φ_{ij} is the (i, j) element of Φ.

Now, we introduce the *logical disjunction* $\oplus$ and the *logical conjunction* $\otimes$ for binary numbers defined as

$$
\begin{aligned}
0 \oplus 0 = 0, \quad & 0 \oplus 1 = 1, \quad 1 \oplus 0 = 1, \quad 1 \oplus 1 = 1 \\
0 \otimes 0 = 0, \quad & 0 \otimes 1 = 0, \quad 1 \otimes 0 = 0, \quad 1 \otimes 1 = 1.
\end{aligned}
\tag{1.6}
$$

Then the relationship between the vector y of test results and x of infection is represented as

$$
\Phi x = y, \tag{1.7}
$$

where the sum and product in this representation are taken as the logical disjunction and logical conjunction. Since the goal is to dramatically reduce the number m of tests, m is much smaller than the number of patients n. In this case, the above equation (1.7) has infinitely many solutions in general (if solutions exist). This means that we cannot uniquely determine the original x only from (1.7). However, if we assume that the number of infected people is much smaller than n, or vector x is *sparse* (i.e., x has very few nonzero elements), then we can formulate group testing as the following optimization problem:

$$
\underset{x \in \{0,1\}^n}{\text{minimize}} \ \|x\|_0 \ \text{subject to} \ \Phi x = y, \tag{1.8}
$$

where $\|x\|_0$ denotes the ℓ^0 norm, the number of nonzero elements in x. We will discuss the ℓ^0 norm in detail in Chapter 2. The optimization problem is a *combinatorial optimization problem* or a *binary optimization problem*, and the computational burden increases exponentially as the size n increases. Dorfman's paper [32] proposes various methods for efficiently running group testing, and an increased interest in this topic has emerged especially with the recent development of sparsity methods. For recent methods, see, for example, [1, 3].

1.3 Optimization with ℓ^1 Norm

As we will see in this book, ℓ^1-norm optimization is one of the most important techniques for sparsity methods as approximation of ℓ^0-norm optimization.

1.3.1 Signal Reconstruction

The first study of optimization with the ℓ^1 norm as a sparsity method is found in the dissertation by Franklin Logan in 1965 [69]. Logan considered the problem of *signal reconstruction* from noisy data. In his dissertation, he showed that ℓ^1-norm

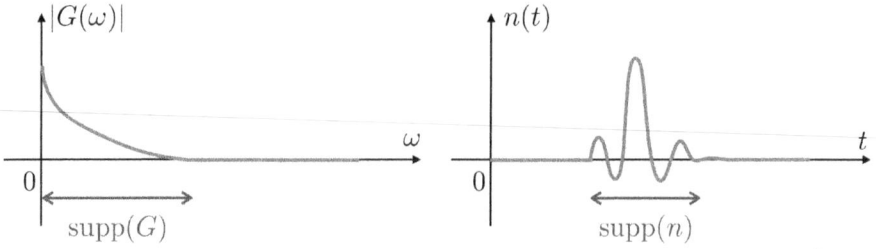

Figure 1.3. Reconstruction from noisy signal $f(t) = g(t) + n(t)$: the Fourier transform $G(\omega)$ of g is band-limited in the frequency domain, and the noise n is localized in the time domain.

minimization completely eliminates the noise when the original signal is band-limited to a certain frequency and the noise is well localized (i.e., sparse) on the time axis. More precisely, if we have noisy observation

$$f(t) = g(t) + n(t), \quad t = 0, 1, 2, \ldots \tag{1.9}$$

where the Fourier transform $G(\omega)$ of g has its support on a low-frequency range, and the support of $n(t)$ is sufficiently short, then the ℓ^1 optimization leads to perfect reconstruction of g from f. This is called *Logan's phenomenon*. Figure 1.3 illustrates the signal assumptions (band limitation and sparsity) in Logan's phenomenon. The sparsity method by ℓ^1-norm minimization was then extended in [31] to signal recovery when the original signal is sparse in the frequency domain (i.e., the original signal may not necessarily be a low-frequency signal).

1.3.2 Geophysics

In the field of *geophysics*, sparsity methods by ℓ^1 optimization have been proposed since the 1970s. The structure of the strata can be estimated by generating artificial earthquakes near the ground surface and observing the reflected waves. This is a method called the *reflection seismic survey*. This is a problem of *system identification* or an *inverse problem*, where the characteristics of the system is estimated from its inputs and outputs. As shown in the left-hand diagram in Figure 1.4, we consider a linear system R with input (wave by the artificial earthquake) $u(t)$ and the output (reflected wave) $y(t)$.

The problem is to find the impulse response $r(t)$ of the system R from the input/output data of $u(t)$ and $y(t)$. In the case of seismic reflection waves, the impulse response $r(t)$ can be assumed to be localized in time (see the right-hand figure in Figure 1.4). That is, the impulse response is *sparse*. From this, the ℓ^1 regularization was proposed to reconstruct the sparse impulse response [23, 100, 109]. These are other early studies that used the idea of sparsity.

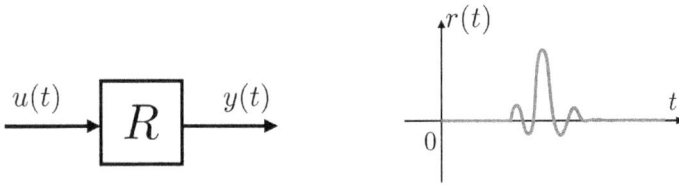

Figure 1.4. Linear system R with input $u(t)$ and output $y(t)$ (left), and its sparse impulse response $r(t)$ (right).

1.3.3 Neural Networks

In the field of *neural networks*, the idea of sparsity has also been investigated. Since the 1980s, Masumi Ishikawa has been proposing a method to avoid overfitting by introducing the ℓ^1 norm regularization into the training of multilayer perceptrons [58]. He proposed to sparsify the coupling weights of the network to avoid overfitting. This is a method of machine learning that takes advantage of the human brain's ability to forget, which is called *structure learning with forgetting*. The method can lead to an explainable structure of multilayer neural networks, which allows people to understand the learning results. Also, the technique of *dropout* in recent deep neural networks is based on a similar idea of sparsity [42, 106].

1.3.4 Statistics

In statistics, the method called *LASSO* (Least Absolute Shrinkage and Selection Operator) is the most famous method with sparsity. Let us consider polynomial curve fitting from given data. If we can set many of the coefficients (parameters) to zero, the terms with zero coefficients will not affect the estimation at all and we can avoid overfitting. Such a method is called a *shrinkage method* in statistics. LASSO, the shrinkage method with ℓ^1 norm regularization, was proposed by Robert Tibshirani in 1996 [110]. The idea of LASSO has been extended to *elastic net regularization* [119] with the sum of the ℓ^1 norm and the squared ℓ^2 norm as a regularized term, and *group LASSO* with the sum of weighted ℓ^2 norms for grouped vectors [116]. We will study LASSO in Chapter 3.

1.3.5 Signal Processing

The first research area where sparsity methods became a hot topic is signal processing. A method called *basis pursuit* with ℓ^1 norm optimization to recover sparse signals was proposed in 1994 by Chen and Donoho [21] at the 28th Asilomar Conference on Signals, Systems, and Computers.[1] In addition, the *total variation*

---

1. Later, this work was published as a journal article [20] with Saunders as a co-author.

denoising, by using the ℓ^1 norm of the difference of a signal was proposed in 1992 by Rudin *et al.* [97]. More recently, Donoho *et al.* proposed a new theory of sensing and recovery called *compressed sensing* [30] in 2006, which is a theoretically refined version of the basis pursuit. In the same year, Candes and Tao also published a paper on this topic [16]. 2006 is the year that the current development of compressed sensing began. Compressed sensing was a topic in the fields of signal processing and information theory at that time. However, the topic is now widely attracting a lot of attention in various research fields including systems and control.

1.4   Sparsity Methods for Systems and Control

Here we describe a brief history of sparsity methods for systems and control to provide a motivation for studying the new research topic.

1.4.1   Minimum Fuel Control and L^1 Optimization

In the field of automatic control, sparsity has been recognized for a long time. An example is the *minimum fuel control*, which is an optimal control that minimizes the L^1 norm of control among feasible controls. The minimum fuel control has been actively discussed in the field of control theory since the early 1960s [2]. At that time, the space race between the United States and the Soviet Union was most heated, and the minimum fuel control has a background in the discussion on how to reduce the fuel consumption of rockets from the earth to the moon, for example. As we will see in Chapter 7, the minimum fuel control is a *bang-off-bang control* that takes ternary values of $\pm u_{\max}$ (the maximum amplitude that the control can produce) or *zero*, under some assumptions. When the control takes a value of zero, the rocket engages in inertial flight, and hence it can reduce fuel consumption during this time. This is why the control is called minimum fuel.

1.4.2   Maximum Hands-off Control

The L^1-optimal minimum-fuel control is shown to be equivalent to the L^0-optimal control (the sparsest control) in [82, 83] under the assumption of non-singularity. The sparse control with minimum L^0 norm is called the *maximum hands-off control*. The mathematical properties of the maximum hands-off control was investigated in [19, 55]. This has also been extended to time-optimal control [57], distributed control [52, 54], continuous control [85], and infinite-dimensional systems [51]. The maximum hands-off control will be discussed in Chapters 8–10.

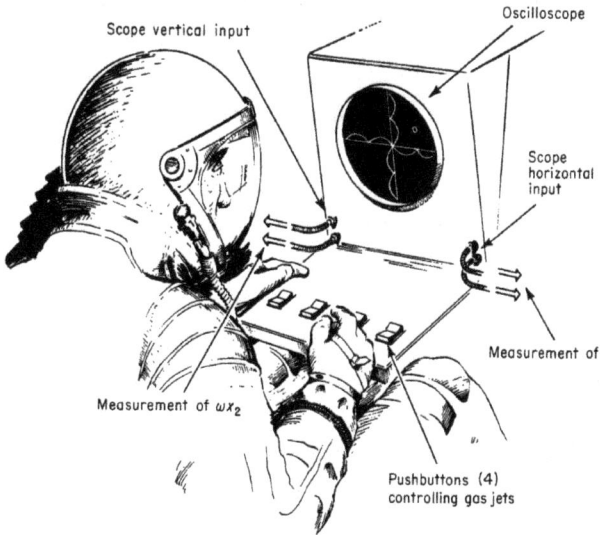

Figure 1.5. The rocket cockpit illustrated in 1960's textbook [2]. The pilot just operates the switches with the observation of the position and velocity of the rocket. This figure is from [2, p. 608, Figure 7-62].

1.4.3 Discrete-valued Control

The ternary property of the minimum fuel control is also understood as *discreteness*. It has been known since the 1960s that certain types of optimal control show such discreteness of control. In fact, the classical textbook [2] states that the discrete-valued control can be implemented as a few switches in the rocket cockpit (see Figure 1.5). Of course, it is obvious that such a simple manual control would be useless and dangerous to fly into space, and that an automatic control with a feed-back mechanism is essential. However, the discrete-valued control expressed only by switching on and off, is very important in recent resource-aware networked control systems, such as the Internet of Things (IoT) or Cyber-Physical Systems (CPS). Discrete-valued control with the idea of sparsity was proposed in [53, 56]. In these papers, the minimization of the *sum of absolute values* (SOAV) of the control to enhance the discreteness. We will study the SOAV control in Section 10.2 of Chapter 10.

1.4.4 Robust Control and Rank Minimization

The optimal control mentioned above requires a complete mathematical model of the controlled object (e.g., a rocket). However, there should be *uncertainties* in the model and parameters in reality, and how to deal with them has been a major challenge in automatic control theory. *Robust control*, a theory of control systems design that takes uncertainty into account, was actively studied in the 1980s, with H^∞ *control theory* being one of the most successful examples (see e.g. [117]). Some basic problems in H^∞ control boil down to the problem of finding a matrix satisfying *linear matrix inequalities* (LMIs) [11, 33]. An LMI is a convex constraint, which can be easily solved using convex optimization (especially the interior point method). However, if you want to control a large-scale and high-dimensional system by a simple and much lower-dimensional controller, or if you need to treat structured uncertainties, the problem becomes LMIs with a matrix rank constraint (or rank minimization), which is much more difficult to solve since the rank constraint is non-convex.[2]

The rank minimization problem is in general described as

$$\underset{X \in \mathbb{R}^{n \times n}}{\text{minimize}} \ \text{rank}(X) \ \text{subject to} \ M(X) + Q \succeq 0 \qquad (1.10)$$

where $M(X)$ is a linear function of X, Q is a matrix, and the inequality $A \succeq B$ means $A - B$ is positive semidefinite. It is easily shown that the matrix rank is the number of non-zero singular values (i.e. the ℓ^0 norm), and hence this is a problem related to sparsity. As mentioned above, the ℓ^0 norm is often approximated by the ℓ^1 norm, and in this case, we minimize the sum of absolute values (i.e. the ℓ^1 norm) of the singular values. This is called the *nuclear norm* and denoted by $\|X\|_*$. That is, the rank minimization problem in (1.10) is approximated to the *nuclear norm minimization*:

$$\underset{X \in \mathbb{R}^{n \times n}}{\text{minimize}} \ \|X\|_* \ \text{subject to} \ M(X) + Q \succeq 0 \qquad (1.11)$$

The pioneering work by Mesbahi and Papavassilopoulos [75] showed the equivalence between (1.10) and (1.11). Interestingly, this was published in 1997 prior to the theory of compressed sensing in 2000s. For the equivalence, they used the property of Z *matrix*, which has not been considered in standard compressed sensing theory. In this book, we do not deal with rank minimization. Readers who are interested in rank minimization may refer to [73].

1.4.5 Resource-aware Control for Networked Control Systems

Sparsity methods have also been applied to *networked control systems*. A networked control system is a feedback control system where the communication between the

---

2.   The rank constraint can be equivalently transformed into a *bilinear matrix inequality* (BMI), which is also difficult to solve.

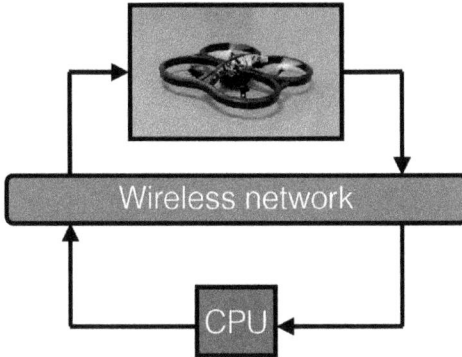

Figure 1.6. Networked control system.

controlled object and the controller is limited. Figure 1.6 shows an example of a networked control system. In this system, sensor data from the drone is sent to the computer (CPU) via a wireless communication network. Based on the information, CPU updates the control values for the attitude, speed, and acceleration of the drone, and returns the control commands to the drone via the network.

For networked control systems, sparsity methods play an important role to realize *resource-aware control* that can significantly reduce the communication and computational burden. In [39, 80, 81, 84, 86, 91], sparse control is proposed by using ℓ^1 norm minimization for discrete-time systems, by which we can reduce the size of control packets that are sent through rate-limited communication networks. These are finite-horizon control and to obtain feedback control, we can adapt the *receding horizon control* or the *model predictive control* formulations.

Minimum actuator placement is also an important sparsity method for resource-aware control. This is to minimize the number of actuators (or control inputs) that achieve a control objective (e.g. controllability). The problem has been discussed in [50, 59, 90, 93, 95, 95, 111].

For state feedback control, the control gain matrix is also sparsified [29, 67, 68, 76]. The obtained feedback controller is sparsely structured and the design should achieve an optimal tradeoff between closed-loop performance and sparsity. See a review paper [61] by Jovanović and Dhingra for detailed discussion on this topic.

Part I

Sparse Representation for Vectors

DOI: 10.1561/9781680837254.ch2

Chapter 2

What is Sparsity?

In this chapter, we explain the notion of *sparsity*, and introduce sparse representation of vectors and functions. The notion introduced in this chapter is important throughout this book, and hence do not omit this chapter.

Key ideas of Chapter 2

- Sparsity of a vector is measured by its ℓ^0 norm.
- In sparse representation, a redundant dictionary of vectors is used.
- In sparse representation, the smallest number of vectors are automatically chosen from a redundant dictionary that represent a given vector (ℓ^0 optimization).
- The exhaustive search to solve ℓ^0 optimization requires computational time that exponentially increases as the problem size increases.

2.1 Redundant Dictionary

Let us consider the three-dimensional vector space $\mathbb{R}^3$. The *standard basis* for $\mathbb{R}^3$ is formed with the following three unit vectors:

$$e_1 = \begin{bmatrix} 1 \\ 0 \\ 0 \end{bmatrix}, \quad e_2 = \begin{bmatrix} 0 \\ 1 \\ 0 \end{bmatrix}, \quad e_3 = \begin{bmatrix} 0 \\ 0 \\ 1 \end{bmatrix}. \tag{2.1}$$

By using this basis, any three-dimensional vector $y \in \mathbb{R}^3$ can be represented as

$$y = \begin{bmatrix} y_1 \\ y_2 \\ y_3 \end{bmatrix} = y_1 e_1 + y_2 e_2 + y_3 e_3. \tag{2.2}$$

In general, if you choose three linearly independent vectors ϕ_1, ϕ_2, and ϕ_3 from $\mathbb{R}^3$, then they form a basis for $\mathbb{R}^3$. That is, for any vector $y \in \mathbb{R}^3$, there exist unique real numbers β_1, β_2, and β_3 such that

$$y = \beta_1 \phi_1 + \beta_2 \phi_2 + \beta_3 \phi_3 \tag{2.3}$$

holds. Moreover, if ϕ_1, ϕ_2, and ϕ_3 are unit vectors and orthogonal to each other, that is,

$$\langle \phi_i, \phi_j \rangle = \phi_j^\top \phi_i = \begin{cases} 1, & i = j \\ 0, & i \neq j \end{cases}, \quad i, j = 1, 2, 3, \tag{2.4}$$

where $\langle \cdot, \cdot \rangle$ is the ℓ^2 inner product (see also Section 2.3). Then ϕ_1, ϕ_2, and ϕ_3 form an *orthonormal basis* for $\mathbb{R}^3$, and the coefficients β_1, β_2, and β_3 can be obtained by the inner product

$$\beta_j = \langle y, \phi_j \rangle = \phi_j^\top y, \quad i = 1, 2, 3. \tag{2.5}$$

Exercise 2.1. How do you obtain the coefficients β_1, β_2, and β_3 in (2.3), when ϕ_1, ϕ_2, and ϕ_3 are linearly independent but they do not form an orthonormal basis?

Let us consider another basis for $\mathbb{R}^3$ with the following three linearly independent vectors:

$$\phi_1 = e_1 + e_2 = \begin{bmatrix} 1 \\ 1 \\ 0 \end{bmatrix}, \quad \phi_2 = e_2 + e_3 = \begin{bmatrix} 0 \\ 1 \\ 1 \end{bmatrix}, \quad \phi_3 = e_3 + e_1 = \begin{bmatrix} 1 \\ 0 \\ 1 \end{bmatrix}. \tag{2.6}$$

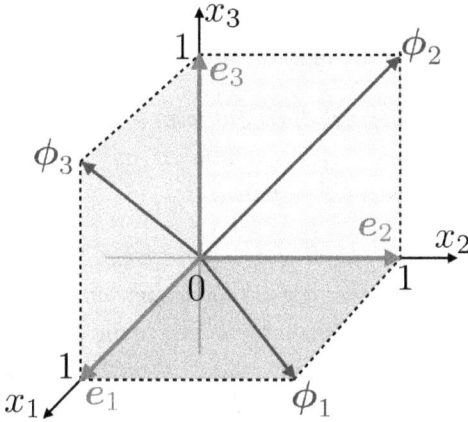

Figure 2.1. 6 vectors $e_1, e_2, e_3, \phi_1, \phi_2, \phi_3$ in $\mathbb{R}^3$.

Combining these with the unit vectors in (2.1), let us form a set of 6 vectors $\{e_1, e_2, e_3, \phi_1, \phi_2, \phi_3\}$. Figure 2.1 shows these 6 vectors. With these vectors, consider the following representation of vector $y \in \mathbb{R}^3$:

$$y = \sum_{i=1}^{3} \alpha_i e_i + \sum_{i=1}^{3} \beta_i \phi_i. \tag{2.7}$$

This is a *redundant* representation, and there are infinitely many solutions for α_i and β_i ($i = 1, 2, 3$) to satisfy (2.7). For example, for $y = [y_1, y_2, y_3]^\top$, we have two solutions

$$(\alpha_1, \alpha_2, \alpha_3, \beta_1, \beta_2, \beta_3) = (y_1, y_2, y_3, 0, 0, 0), \tag{2.8}$$

and

$$(\alpha_1, \alpha_2, \alpha_3, \beta_1, \beta_2, \beta_3) = (-y_3, -y_1, -y_2, y_1, y_2, y_3). \tag{2.9}$$

Now, let us consider a situation where the cost to keep the values of the non-zero coefficients is very expensive due to an expensive memory device for example. Then we want to minimize the number of non-zero coefficients to reduce the cost. Let us consider a vector $y \in \mathbb{R}^3$ on the plane spanned by e_1 and ϕ_2. For this vector, we have the following solution:

$$y = \alpha_1 e_1 + \beta_2 \phi_2, \tag{2.10}$$

This expression has smaller number of non-zero coefficients than (2.8). This is a trivial example, and the cost of (2.10) is almost the same as that of (2.8). However, if we can find just 10^2 non-zero coefficients for a 10^6 (one million) dimensional vector, the cost will be dramatically reduced. Such a technology is often called *data compression*, which is one of the biggest motivations of *sparse representation*.

Example 2.2. *The four cardinal directions form a redundant system to represent a direction in $\mathbb{R}^2$. We say for example "Go southwest" not "Go minus-north-minus-east" although the two are mathematically equivalent.*

Example 2.3. *Imagine that you need to explain what an elephant is to a foreigner, who cannot speak English but has a small dictionary with 3000 words but the word "elephant." You might say "Elephant is the largest living land animal that has a long nose, many of them live in African savanna..." then the foreigner will ask "What is savanna?" since the word is not in the foreigner's dictionary. But if the foreigner has a large dictionary that has more than 1 million words, you just say "That is an elephant." Some English teachers say you need to memorize only these 3000 words for conversation, but actually 3000 words are not enough at all for* simple *expression.*

Let us formulate this problem of sparse representation in a general form. Let us consider m-dimensional vector space $\mathbb{R}^m$, and a set of vectors $\{\phi_1, \phi_2, \ldots, \phi_n\}$ in $\mathbb{R}^m$, where $m < n$. For a given vector $y \in \mathbb{R}^m$, we find coefficients $\alpha_1, \alpha_2, \ldots, \alpha_n$ such that

$$y = \sum_{i=1}^{n} \alpha_i \phi_i. \tag{2.11}$$

We assume that m vectors in $\{\phi_1, \phi_2, \ldots, \phi_n\}$ are linearly independent. We call such a set of vectors $\{\phi_1, \phi_2, \ldots, \phi_n\}$ a *dictionary* (recall Example 2.3), and the elements $\phi_1, \phi_2, \ldots, \phi_n$ *atoms*.[1] Note that the size n of the dictionary is larger than the size m of vector y. We call such a dictionary a *redundant dictionary*, or *over-complete dictionary*.

Define a matrix Φ and a vector x as

$$\Phi \triangleq \begin{bmatrix} \phi_1 & \phi_2 & \cdots & \phi_n \end{bmatrix} \in \mathbb{R}^{m \times n}, \quad x \triangleq \begin{bmatrix} \alpha_1 \\ \alpha_2 \\ \vdots \\ \alpha_n \end{bmatrix} \in \mathbb{R}^n. \tag{2.12}$$

Then, the equation (2.11) can be equivalently written as

$$\Phi x = y. \tag{2.13}$$

---

1.   We do not call them *words*.

The matrix Φ is called a *dictionary matrix*, or a *measurement matrix*. Since the dictionary is redundant, the matrix Φ is a *fat* matrix, that is, the number of columns is larger than the number of rows. Our problem is now described as follows:

Problem 2.4 (Sparse Representation). *Given a vector $y \in \mathbb{R}^m$ and a dictionary* $\{\phi_1, \phi_2, \ldots, \phi_n\}$. *Find the simplest representation of y that satisfies* (2.13).

In the next section, we discuss this problem with a fat matrix.

2.2   Underdetermined Systems

Let us consider the following system of linear equations with unknowns x_1, x_2, and x_3:

$$\begin{aligned} x_1 + x_2 + x_3 &= 3 \\ x_1 - x_3 &= 0 \end{aligned} \tag{2.14}$$

Now there are three unknowns and two equations, and it is easily seen that there are infinitely many solutions. To represent all solutions, we use parametrization. All solutions of (2.14) are parametrized as

$$x_1 = t, \quad x_2 = -2t + 3, \quad x_3 = t, \tag{2.15}$$

where $t \in \mathbb{R}$ is a parameter. We call such a system of equations an *underdetermined system*, where the number of unknowns is larger than the number of equations.

An underdetermined system is something like insufficient proofs for a detective to determine one among many suspects. For a detective, say Conan Edogawa,[2] the two proofs (equations) in (2.14) are insufficient and he should seek one more proof to reveal the unique solution of the case. Thanks to his investigation, a proof was found, which said *"the criminal is the smallest one among the suspects."* This is actually a conclusive proof that can choose just one suspect. Let us find the smallest solution among the candidates in (2.15). We use the ℓ^2 norm as a measure of the size, and we find the smallest ℓ^2-norm solution as follows. First, from (2.15), we have

$$\begin{aligned} \|x\|_2^2 &= x_1^2 + x_2^2 + x_3^2 \\ &= t^2 + (-2t + 3)^2 + t^2 \\ &= 6(t - 1)^2 + 3. \end{aligned} \tag{2.16}$$

Then we can choose $t = 1$, and from (2.15), the solution is uniquely chosen as $(x_1, x_2, x_3) = (1, 1, 1)$. *Case closed.*

---

2.   See: https://en.wikipedia.org/wiki/Case_Closed

Let us generalize the above discussion. We consider a system of linear equations in a matrix form as

$$\Phi x = y. \tag{2.17}$$

For example, the system in (2.14) can be represented in the matrix form (2.17) with

$$\Phi = \begin{bmatrix} 1 & 1 & 1 \\ 1 & 0 & -1 \end{bmatrix}, \quad x = \begin{bmatrix} x_1 \\ x_2 \\ x_3 \end{bmatrix}, \quad y = \begin{bmatrix} 3 \\ 0 \end{bmatrix}. \tag{2.18}$$

We assume the size of matrix Φ is $m \times n$ where $m < n$, that is, we consider an underdetermined system of equations. We also assume that there are m column vectors in $\{\phi_1, \ldots, \phi_n\}$ that are linearly independent. In other words, we assume Φ has *full row rank*. Note that a matrix $\Phi \in \mathbb{R}^{m \times n}$ is said to have full row rank if Φ is *surjective*, or

$$\text{rank}(\Phi) = m. \tag{2.19}$$

If $\text{rank}(\Phi) < m$, then there exist redundant linear equations (i.e. there is at least one equation that is a linear combination of other equations). For example, the following system of equations

$$\begin{aligned} x_1 + x_2 + x_3 &= 3 \\ x_1 - x_3 &= 0 \\ 2x_1 + x_2 &= 3 \end{aligned} \tag{2.20}$$

is redundant and the rank is $2 < 3$. We here assume such redundancy should be eliminated beforehand.

If Φ has full row rank, then for any vector $y \in \mathbb{R}^m$, there exists at least one solution x that satisfies the linear equation (2.17). Let denote by x_0 a *particular solution* of (2.17).

Define the *kernel* (or *null space*) of matrix Φ by

$$\ker(\Phi) \triangleq \{x \in \mathbb{R}^n : \Phi x = 0\}. \tag{2.21}$$

Note that $\ker(\Phi)$ is a linear subspace in $\mathbb{R}^n$, that is, if $x_1, x_2 \in \ker(\Phi)$, then $a_1 x_1 + a_2 x_2 \in \ker(\Phi)$ for any $a_1, a_2, \in \mathbb{R}$.

Then, we introduce the *dimension theorem* in linear algebra.

Theorem 2.5 (dimension theorem). *For any matrix $\Phi \in \mathbb{R}^{m \times n}$,*

$$\text{rank}(\Phi) + \dim \ker(\Phi) = n \tag{2.22}$$

holds.

From the dimension theorem, the dimension of $\ker(\Phi)$ is $n - m$. Since $n > m$, the kernel, which is a linear subspace in $\mathbb{R}^n$, has at least one dimension. That is, there exist infinitely many vectors in $\ker(\Phi)$. Then, all solutions of the linear equation (2.17) can be represented by the sum of a particular solution x_0 and a *free parameter* $z \in \ker(\Phi)$, that is,

$$x = x_0 + z, \quad z \in \ker(\Phi). \tag{2.23}$$

From this it follows that there exist infinitely many solutions of (2.17).

Exercise 2.6. Show that the vector x in (2.23) is the solution of the equation (2.17).

The problem of sparse representation (Problem 2.4) is to find a solution x of (2.17) that has the simplest representation, or the smallest number of non-zero elements. Let us consider this problem more precisely in the next section.

2.3 The ℓ^0 Norm

We here review the notion of a norm in a finite-dimensional vector space, and then introduce the ℓ^0 norm that defines the sparsity of a vector.

First, let us recall the definition of a norm in $\mathbb{R}^n$.

Definition 2.7. *A norm* $\|x\| : \mathbb{R}^n \to [0, \infty)$ *is a nonnegative function that satisfies the following properties:*

1. *For any vector $x \in \mathbb{R}^n$ and any number $\alpha \in \mathbb{R}$, $\|\alpha x\| = |\alpha| \|x\|$.*
2. *For any $x, y \in \mathbb{R}^n$, $\|x + y\| \leq \|x\| + \|y\|$.*
3. *$\|x\| = 0 \iff x = 0$.*

A well-known norm in $\mathbb{R}^n$ is the ℓ^2 norm (or the *Euclidean norm*). For a vector $x = [x_1, x_2, \ldots, x_n]^\top \in \mathbb{R}^n$, the ℓ^2 norm is defined by

$$\|x\|_2 \triangleq \sqrt{x_1^2 + x_2^2 + \cdots + x_n^2}. \tag{2.24}$$

The ℓ^2 norm is also given by

$$\|x\|_2 = \sqrt{\langle x, x \rangle}, \tag{2.25}$$

where $\langle \cdot, \cdot \rangle$ is the ℓ^2 *inner product* (or *Euclidean inner product*) in $\mathbb{R}^n$, defined by

$$\langle x, y \rangle \triangleq y^\top x = \sum_{i=1}^{n} x_i y_i. \tag{2.26}$$

Exercise 2.8. Confirm the ℓ^2 norm $\|x\|_2$ defined in (2.24) satisfies the three properties in Definition 2.7.

Not only the ℓ^2 norm, we can define (infinitely) many norms for $\mathbb{R}^n$. A generalization of the ℓ^2 norm in (2.24) is the ℓ^p *norm* with $p \in [1, \infty)$, defined by

$$\|x\|_p \triangleq \left(\sum_{i=1}^{n} |x_i|^p \right)^{1/p}. \tag{2.27}$$

The most important norm in this book is the ℓ^1 *norm* with $p = 1$ in (2.27). The ℓ^1 norm is described as the sum of the absolute values of the elements in a vector, that is,

$$\|x\|_1 = \sum_{i=1}^{n} |x_i|. \tag{2.28}$$

The limit of (2.27) as $p \to \infty$ is called the ℓ^∞ *norm* (or the *maximum norm*), defined by

$$\|x\|_\infty \triangleq \max_{i=1,2,\ldots,n} |x_i|. \tag{2.29}$$

Exercise 2.9. Prove that for any $x \in \mathbb{R}^n$,

$$\|x\|_\infty = \lim_{p \to \infty} \|x\|_p. \tag{2.30}$$

Figure 2.2 shows the contour curves that satisfy $\|x\|_p = 1$ for $p = 1, 2$, and ∞ in $\mathbb{R}^2$. The contour of the ℓ^2 norm is a unit circle centered at the origin. The

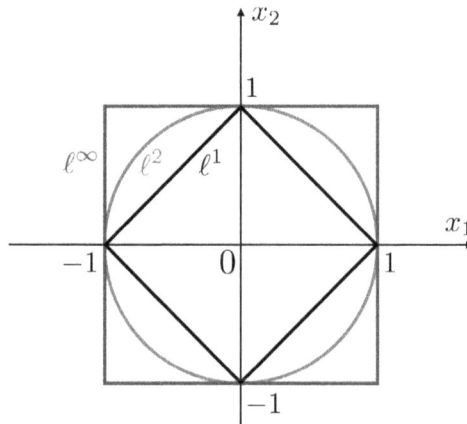

Figure 2.2. Contour curves ($\|x\|_p = 1$) of ℓ^1, ℓ^2, ℓ^∞ norms.

contour of the ℓ^∞ norm is a unit square centered at the origin, and touches the ℓ^2 circle at $(1, 0)$, $(0, 1)$, $(-1, 0)$, and $(0, -1)$. The shape of the contour of the ℓ^1 norm is very important for sparse representation. This diamond-shaped contour has four corners on the x_1 and x_2 axes. This property gives an intuitive explanation of the relation between ℓ^1 norm and sparsity (see Section 3.2).

Now, let us define the ℓ^0 norm. Consider a vector $x = [x_1, x_2, \ldots, x_n]^\top \in \mathbb{R}^n$. Define the *support* of x by

$$\text{supp}(x) \triangleq \{i \in \{1, 2, \ldots, n\} : x_i \neq 0\}. \tag{2.31}$$

The support of x is the set of indices on which the elements of x are nonzero. By using the support, the ℓ^0 *norm* is defined by

$$\|x\|_0 \triangleq \#(\text{supp}(x)), \tag{2.32}$$

where $\#(\text{supp}(x))$ is the number of elements in finite set $\text{supp}(x)$. Namely, the ℓ^0 norm counts the number of nonzero elements in x.

It is notable that the ℓ^0 norm does not satisfy the first property in Definition 2.7. For example, a nonzero vector $x \in \mathbb{R}^n$ has the same ℓ^0 norm as $2x$. This implies that

$$\|2x\|_0 = \|x\|_0 \neq 2\|x\|_0, \tag{2.33}$$

whenever $x \neq 0$. Strictly speaking, the ℓ^0 norm is not a norm, and hence we sometimes call it as ℓ^0 *pseudo-norm* or *cardinality*. However, we use the term "ℓ^0 norm" as often used in the literature. Note that by definition, the second and third properties in Definition 2.7 hold, that is,

$$\|x + y\|_0 \leq \|x\|_0 + \|y\|_0 \tag{2.34}$$

and

$$\|x\|_0 = 0 \iff x = 0. \tag{2.35}$$

Finally, we define the sparsity of a vector by using the ℓ^0 norm. A vector $x \in \mathbb{R}^n$ is said to be *sparse* if the ℓ^0 norm $\|x\|_0$ is sufficiently small compared to the dimension n. The notion of sparsity is important in this book.

Exercise 2.10. Prove that for any $x, y \in \mathbb{R}^n$,

$$\|x + y\|_0 \leq \|x\|_0 + \|y\|_0 \tag{2.36}$$

holds.

Exercise 2.11. Let $x, y \in \mathbb{R}^n$. When does the following equality hold?

$$\|x + y\|_0 = \|x\|_0 + \|y\|_0. \tag{2.37}$$

The problem of sparse representation Problem 2.4 is finding the sparsest solution among infinitely many solutions of the linear equation in (2.17). This problem is mathematically formulated by using the ℓ^0 norm introduced above. That is, we seek the smallest ℓ^0-norm solution for (2.17). This is formulated as a mathematical optimization problem as follows:

Problem 2.12 (Sparse representation). *Given a vector* $y \in \mathbb{R}^m$ *and a full-row-rank matrix* $\Phi \in \mathbb{R}^{m \times n}$ *with* $m < n$. *Find the optimizer* x^* *of the optimization problem:*

$$\underset{x \in \mathbb{R}^n}{\text{minimize}} \ \|x\|_0 \ \text{subject to} \ \Phi x = y. \tag{2.38}$$

We call this the ℓ^0 *optimization.*

2.4 Exhaustive Search

In this section, we show a direct method to solve the ℓ^0 optimization problem (2.38), called an *exhaustive search* (or *brute-force search*). Let $\phi_i \in \mathbb{R}^m$ ($i = 1, 2, \ldots, n$) denote the i-th column vector in matrix Φ, that is,

$$\Phi \triangleq \begin{bmatrix} \phi_1 & \phi_2 & \cdots & \phi_n \end{bmatrix} \in \mathbb{R}^{m \times n}. \tag{2.39}$$

The following shows the procedure of the exhaustive search for (2.38).

1. If $y = 0$, then output $x^* = 0$ as the optimal solution and quit. Otherwise, proceed to the next step.
2. Find a vector x with $\|x\|_0 = 1$ that satisfies the equation $y = \Phi x$. That is, set

$$x_1 \triangleq \begin{bmatrix} x_1 \\ 0 \\ \vdots \\ 0 \end{bmatrix}, \quad x_2 \triangleq \begin{bmatrix} 0 \\ x_2 \\ 0 \\ \vdots \\ 0 \end{bmatrix}, \ldots, \quad x_n \triangleq \begin{bmatrix} 0 \\ \vdots \\ 0 \\ x_n \end{bmatrix} \tag{2.40}$$

and search $x_i \in \mathbb{R}$ ($i = 1, 2, \ldots, n$) that satisfies

$$y = \Phi x_i = x_i \phi_i. \tag{2.41}$$

If a solution exists for some i, output $x^* = x_i$ as the solution and quit. Otherwise, proceed to the next step.

3. Find a vector x with $\|x\|_0 = 2$ that satisfies the equation $y = \Phi x$. That is, set

$$x_{1,2} \triangleq \begin{bmatrix} x_1 \\ x_2 \\ 0 \\ \vdots \\ 0 \end{bmatrix}, \quad x_{1,3} \triangleq \begin{bmatrix} x_1 \\ 0 \\ x_3 \\ 0 \\ \vdots \\ 0 \end{bmatrix}, \dots, \quad x_{n-1,n} \triangleq \begin{bmatrix} 0 \\ \vdots \\ 0 \\ 0 \\ x_{n-1} \\ x_n \end{bmatrix} \quad (2.42)$$

and search $x_i, x_j \in \mathbb{R}$ $(i, j = 1, 2, \dots, n)$ that satisfies

$$y = \Phi x_{i,j} = x_i \phi_i + x_j \phi_j. \quad (2.43)$$

If a solution exists for some i, j, then output $x^* = x_{i,j}$ and quit. Otherwise, proceed to the next step.

4. Do similar procedures for $\|x\|_0 = k, k = 3, 4, \dots, m$.

By this exhaustive search, you can obtain the optimal solution x^* (if it exists) with finite number of steps (the worst case is $k = m$, where the optimal value is $\|x^*\|_0 = m$).

Next, we investigate the exhaustive search in detail. For a vector $x = [x_1, x_2, \dots, x_n]^\top$ and an index set $S \subset \{1, 2, \dots, n\}$, we denote by $x_S \in \mathbb{R}^{\#(S)}$ the restriction of x to the indices in S, where $\#(S)$ is the number of elements in S. For example, for $x = [x_1, x_2, x_3, x_4, x_5, x_6]^\top$ and $S = \{1, 2, 5\}$, we have

$$x_S = \begin{bmatrix} x_1 \\ x_2 \\ x_5 \end{bmatrix} \in \mathbb{R}^3. \quad (2.44)$$

More generally, for $x = [x_1, x_2, \dots, x_n]^\top$ and the index set

$$S = \{i_1, i_2, \dots, i_k\}, \quad k \in \{1, 2, \dots, n\}, \quad (2.45)$$

where $1 \le i_1 < i_2 < \cdots < i_k \le n$, we have

$$x_S = \begin{bmatrix} x_{i_1} \\ x_{i_2} \\ \vdots \\ x_{i_k} \end{bmatrix} \in \mathbb{R}^k. \quad (2.46)$$

Also, for matrix $\Phi = [\phi_1, \phi_2, \ldots, \phi_n] \in \mathbb{R}^{m \times n}$ with $\phi_i \in \mathbb{R}^m$, $i = 1, 2, \ldots, n$ and the index set in (2.45), we define

$$\Phi_S = [\phi_{i_1}, \phi_{i_2}, \ldots, \phi_{i_k}] \in \mathbb{R}^{m \times k}. \tag{2.47}$$

Using this notation, we can formulate the exhaustive search algorithm for the ℓ^0 optimization (2.38) as follows: First check if $y = 0$. In this case, the solution is $x^* = 0$. Otherwise, take each subset S of the index set $\{1, 2, \ldots, n\}$ from $\#(S) = 1$ to $\#(S) = m$, and solve the following equation

$$y = \Phi_S x_S. \tag{2.48}$$

If there is a solution of (2.48), then using the solution $x_S = [x_{i_1}, \ldots, x_{i_k}]^\top$, set $x^* = [x_1^*, x_2^*, \ldots, x_n^*]^\top$ where

$$x_i^* = \begin{cases} x_i, & i \in S, \\ 0 & i \notin S. \end{cases} \tag{2.49}$$

This is the sparsest solution and we have $\|x^*\|_0 = k$. We summarize the exhaustive search algorithm.

Exhaustive search algorithm for ℓ^0 optimization (2.38)

1. If $y = 0$ then output $x^* = 0$ and quit. Otherwise, proceed to the next step.
2. $k := 1$
3. For each subset $S \subset \{1, 2, \ldots, n\}$ with $\#(S) = k$, do
   - Check if equation $y = \Phi_S x_S$ has a solution.
   - If it exists, output x^* as in (2.49) and quit.
4. $k := k + 1$. Return to 3.

We should notice that with the exhaustive search method the computation time to find a solution grows exponentially with problem size (i.e. m). For example, in image processing, the dimension becomes millions or larger, and the exhaustive search is not useful at all.

Exercise 2.13. For the optimization problem (2.38) of size m, compute the number of iterations at the worst case where the optimal solution x^* has its ℓ^0 norm as $\|x\|_0 = m$. Then, let $m = 100$. Suppose that you can use a supercomputer that can do one iteration of the exhaustive search algorithm in 10^{-15} seconds. Compute the total time needed to do the exhaustive search at the worst case.

The above problem is also known as *combinatorial optimization*, which is in general hard to solve for large-scale problems. In the following chapters, we investigate efficient algorithms for such a hard problem of sparse optimization.

2.5 Sparse Representation for Functions

In this section, we discuss sparse representation for functions.

Let us consider the function space $L^2(0, T)$, the space of all square integrable functions on $(0, T)$ in the sense of Lebesgue. That is, for any $f \in L^2(0, T)$, the L^2 norm is finite:

$$\|f\|_2 \triangleq \int_0^T |f(t)|^2 dt < \infty. \tag{2.50}$$

In this space, we can define the L^2 inner product

$$\langle f, g \rangle \triangleq \int_0^T f(t)\overline{g(t)}dt, \tag{2.51}$$

where $\overline{g(t)}$ is the complex conjugate of $g(t)$. It is well-known that under the L^2 inner product, the space L^2 becomes a Hilbert space.

Then let us consider an orthonormal basis $\{\phi_i : i \in \mathbb{Z}\}$ in $L^2(0, T)$ that satisfies

$$\langle \phi_i, \phi_j \rangle = \delta_{ij} = \begin{cases} 1, & \text{if } i = j, \\ 0, & \text{otherwise.} \end{cases} \tag{2.52}$$

Then, for any function $f \in L^2(0, T)$, there exist a complex sequence $\{\alpha_i : i \in \mathbb{Z}\}$ such that

$$f = \sum_{i=-\infty}^{\infty} \alpha_i \phi_i, \tag{2.53}$$

where the convergence is in the sense of L^2, that is,

$$\left\| f - \sum_{i=-N}^{N} \alpha_i \phi_i \right\|_2 \to 0, \tag{2.54}$$

as $N \to \infty$. The representation (2.53) is called *Fourier series*[3] of f. Given f and $\{\phi_i\}$, the coefficients are obtained by the inner product

$$\alpha_i = \langle f, \phi_i \rangle = \int_0^T f(t)\overline{\phi_i(t)}dt. \qquad (2.55)$$

Exercise 2.14. Prove that (2.55) holds.

A standard basis for $L^2(0, T)$ is the *Fourier basis* defined by

$$\phi_i(t) \triangleq \frac{1}{\sqrt{T}}e^{j\omega_i t}, \quad i \in \mathbb{Z} \qquad (2.56)$$

where $j = \sqrt{-1}$ and $\omega_i = 2\pi i / T$. With this basis, the coefficients in (2.55) are given as

$$\alpha_i = \frac{1}{\sqrt{T}} \int_0^T f(t)\overline{e^{j\omega_i t}}dt = \frac{1}{\sqrt{T}} \int_0^T f(t)e^{-j\omega_i t}dt. \qquad (2.57)$$

For a sufficiently smooth function, the Fourier basis gives a good solution to represent the function with a finite number of coefficients by *truncation*. That is, we approximate function f as

$$f_N = \sum_{i=-N}^N \alpha_i \phi_i, \quad \alpha_i = \frac{1}{\sqrt{T}} \int_0^T f(t)e^{-j\omega_i t}dt. \qquad (2.58)$$

Actually, this is optimal in the sense that f_N minimizes the L^2 error

$$\mathcal{E}_N(\beta_{-N}, \ldots, \beta_N) \triangleq \left\| f - \sum_{i=-N}^N \beta_i \phi_i \right\|_2 \qquad (2.59)$$

among all coefficients $\{\beta_{-N}, \ldots, \beta_N\}$.

Now, let us consider a rectangular function on $L^2(0, 1)$ defined by

$$f(t) = \begin{cases} 1, & t \in (0, 1/2), \\ -1, & t \in [1/2, 1). \end{cases} \qquad (2.60)$$

Figure 2.3 (left) shows this function. We can see that this function is discontinuous. The Fourier coefficients of this function can be easily computed using (2.57).

---

3. This is also called as *generalized* Fourier series. Then, with the standard Fourier basis in (2.56), the series in (2.53) is called the Fourier series.

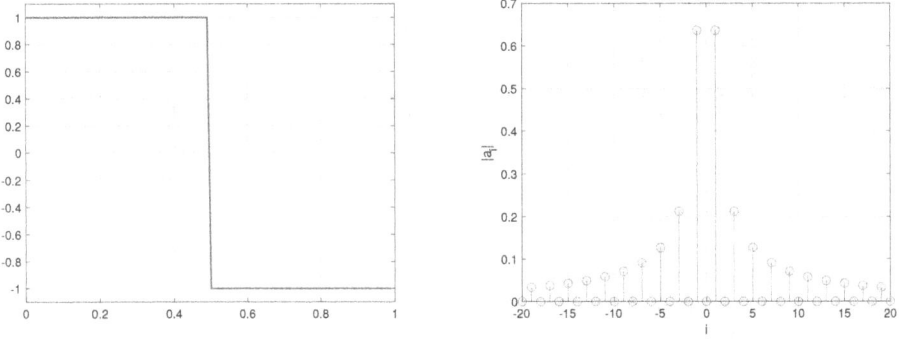

Figure 2.3. Discontinuous rectangular function $f(t)$ (left) and absolute values of its Fourier coefficients (right).

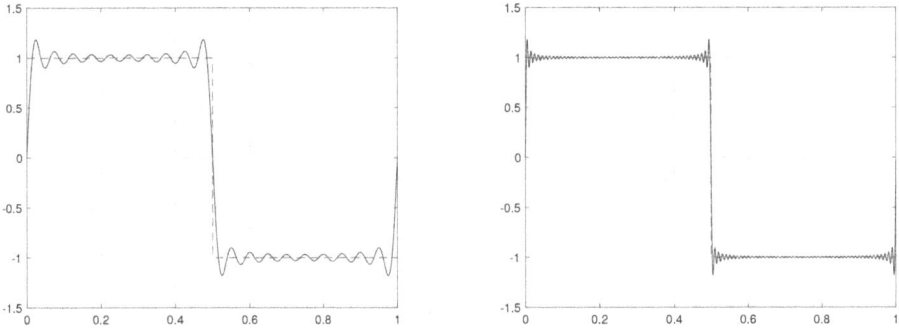

Figure 2.4. Truncated Fourier series $f_N(t)$ with $N = 20$ (left) and $N = 100$ (right).

In fact, we have

$$
\alpha_i = \begin{cases} -\frac{2j}{\pi i}, & \text{if } i \text{ is odd,} \\ 0, & \text{otherwise.} \end{cases} \tag{2.61}
$$

Exercise 2.15. Show that the Fourier coefficients of the rectangular function in (2.60) are given by (2.61).

Figure 2.3 (right) shows the absolute values of the coefficients with $i = -20$ to 20. We can see that the coefficients converge to zero as i goes to $\pm\infty$. Actually, from (2.61), the coefficient sequence $\{\alpha_i\}$ converges to zero as $|i| \rightarrow \infty$ with convergence rate $O(1/i)$.

This fact suggests us to truncate the coefficients with N to obtain the approximant $f_N(t)$ in (2.58). Figure 2.4 shows the truncated Fourier series $f_N(t)$ in (2.58).

The left figure is $f_N(t)$ with $N = 20$. We see oscillations around the edges of the rectangular function. When we increase N to $N = 100$, we obtain another

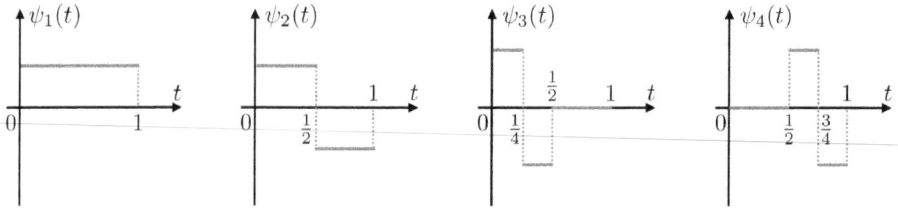

Figure 2.5. Haar functions ψ_1, ψ_2, ψ_3, and ψ_4.

oscillative function in the right figure of Figure 2.4. This oscillation never disappears around the edges for arbitrarily large but finite N. This is called *Gibbs phenomenon*. To store the information of the rectangular function as Fourier coefficients, you cannot truncate it but store all of the coefficients.

Let us consider another orthonormal basis in $L^2(0, 1)$ called *Haar basis* defined by *Haar functions*

$$\psi_1(t) \triangleq 1, \tag{2.62}$$

and for $i = 2^m + k, k = 1, 2, \ldots, 2^m, m = 0, 1, 2, \ldots,$

$$\psi_i(t) \triangleq \begin{cases} \sqrt{2^m}, & t \in \left[2^{-m}(k-1), 2^{-m-1}(2k-1)\right), \\ -\sqrt{2^m}, & t \in \left[2^{-m-1}(2k-1), 2^{-m}k\right), \\ 0, & \text{otherwise.} \end{cases} \tag{2.63}$$

Figure 2.5 shows Haar functions ψ_1, ψ_2, ψ_3, and ψ_4.

Then, if we adopt a *redundant* dictionary of bases consists of the Fourier basis in (2.56) and the Haar basis. From this dictionary, we can simply represent the rectangular function in (2.60) as

$$f(t) = \psi_2(t). \tag{2.64}$$

That is, we need to store just one coefficient under the redundant basis. This is the motivation to use a redundant dictionary and to obtain sparse representation for functions. As shown above, sparse representation of functions is to sparsify the coefficients in the Fourier series of a given function.

2.6 Further Readings

The notion of redundant dictionary and sparse optimization described in this chapter is fundamental and important in this book. The redundant representation of vectors is related to *frames* and *wavelets*, for which readers can refer to nice books by Strang and Nguyen [107] and by Mallat [72]. For fundamental theory of vector spaces, called functional analysis, including norms, inner products, orthonormal bases, and Fourier series, I recommend books by Young [115] and by Yamamoto [114], which are written for scientists and engineers.

DOI: 10.1561/9781680837254.ch3

Chapter 3

Curve Fitting and Sparse Optimization

In this chapter, we study *curve fitting* to obtain a curve, or a function, from given data, and how sparse optimization effectively works for this problem.

> **Key ideas of Chapter 3**
>
> - Curve fitting is formulated as an optimization problem to choose one solution among (infinitely many) candidates.
> - Regularization is used for avoiding overfitting.
> - Sparse optimization is reduced to ℓ^1 optimization, which is convex and efficiently solved by numerical optimization.

3.1 Least Squares and Regularization

We begin with the least squares and regularization with simple examples.

3.1.1 Underdetermined System and Minimum ℓ^2-Norm Solution

Let us consider the linear equation

$$\Phi x = y, \tag{3.1}$$

where $y \in \mathbb{R}^m$ is a given vector, $\Phi \in \mathbb{R}^{m \times n}$ is a given matrix, and $x \in \mathbb{R}^n$ is an unknown vector. We here assume $m < n$, and Φ has full row rank, that is,

$$\mathrm{rank}(\Phi) = m. \tag{3.2}$$

Under these assumptions, there exist infinitely many solutions of the equation (3.1). Let us find the smallest ℓ^2-norm solution among them. This is formulated as an optimization problem

$$\underset{x \in \mathbb{R}^n}{\mathrm{minimize}} \ \frac{1}{2}\|x\|_2^2 \quad \text{subject to } \Phi x = y. \tag{3.3}$$

We call this problem the ℓ^2 *optimization problem*, and the solution the *minimum* ℓ^2-norm solution.

To solve this problem, we can use the *method of Lagrange multipliers*. First, we define the *Lagrange function*, or simply *Lagrangian*, of the optimization problem (3.3) by

$$L(x, \lambda) = \frac{1}{2}x^\top x + \lambda^\top (\Phi x - y). \tag{3.4}$$

The variable $\lambda \in \mathbb{R}^m$ is called the *Lagrange multiplier*.

Then, we can obtain the optimal solution of (3.3) by finding the stationary point (x^*, λ^*) of the Lagrange function L. By differentiating L by the variable x, we have

$$\frac{\partial L}{\partial x} = \frac{\partial}{\partial x}\left(\frac{1}{2}x^\top x + \lambda^\top \Phi x\right) = x + \Phi^\top \lambda. \tag{3.5}$$

It follows that the stationary point (x^*, λ^*) satisfies

$$x^* + \Phi^\top \lambda^* = 0. \tag{3.6}$$

Then differentiating L by λ gives

$$\frac{\partial L}{\partial \lambda} = \Phi x - y, \tag{3.7}$$

and hence

$$\Phi x^* - y = 0. \tag{3.8}$$

From this and (3.6), we have

$$-\Phi \Phi^\top \lambda^* = y. \tag{3.9}$$

Since Φ has full row rank, the matrix $\Phi\Phi^\top$ is nonsingular and has its inverse. Therefore, from (3.9) we have

$$\lambda^* = -(\Phi\Phi^\top)^{-1}y. \tag{3.10}$$

Assigning this to (3.6) gives the minimum ℓ^2-norm solution x^* as

$$x^* = \Phi^\top(\Phi\Phi^\top)^{-1}y. \tag{3.11}$$

In summary, if we are given a full-row-rank matrix Φ and a vector y, we can compute the minimum ℓ^2-norm solution by the formula (3.11).

Exercise 3.1. Find the minimum ℓ^2-norm solution of the following equation with unknowns x_1 and x_2:

$$a_1x_1 + a_2x_2 = 1, \tag{3.12}$$

where a_1 and a_2 are nonzero real numbers.

Exercise 3.2. Let $\Phi \in \mathbb{R}^{m \times n}$. Prove that $\Phi\Phi^\top$ is invertible if Φ has full row rank.

3.1.2 Regression and Least Squares

Suppose we are given two-dimensional data

$$\mathcal{D} = \{(t_1, y_1), (t_2, y_2), \ldots, (t_m, y_m)\}. \tag{3.13}$$

Let us consider a polynomial of order $n - 1$,

$$y = f(t) = a_{n-1}t^{n-1} + a_{n-2}t^{n-2} + \cdots + a_1t + a_0. \tag{3.14}$$

Curve fitting is to find coefficients $a_0, a_1, \ldots, a_{n-1}$ with which the polynomial curve has the best fit to the m-point data (see e.g. Figure 3.1). For example, $t_1, t_2, \ldots, t_m$ are sampling instants, and $y_1, y_2, \ldots, y_m$ are temperature data from a sensor at a portion. From these data, we often want to know the curve behind the data. We call such data analysis the *regression analysis* or *polynomial curve fitting*.

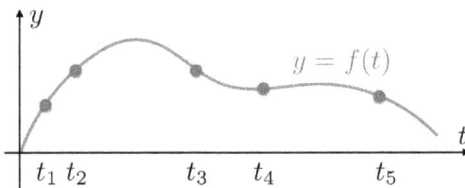

Figure 3.1. Interpolating polynomial.

First, we consider an *interpolating polynomial* that interpolates the given data as shown in Figure 3.1. The polynomial curve (3.14) goes through the data points (3.13), and hence we have m linear equations with unknowns $a_{n-1}, a_{n-2}, \ldots, a_1, a_0$:

$$
\begin{aligned}
a_{n-1}t_1^{n-1} + a_{n-2}t_1^{n-2} + \cdots + a_1 t_1 + a_0 &= y_1, \\
a_{n-1}t_2^{n-1} + a_{n-2}t_2^{n-2} + \cdots + a_1 t_2 + a_0 &= y_2, \\
&\cdots \\
a_{n-1}t_m^{n-1} + a_{n-2}t_m^{n-2} + \cdots + a_1 t_m + a_0 &= y_m.
\end{aligned}
\tag{3.15}
$$

Define a matrix

$$
\Phi \triangleq
\begin{bmatrix}
t_1^{n-1} & t_1^{n-2} & \cdots & t_1 & 1 \\
t_2^{n-1} & t_2^{n-2} & \cdots & t_2 & 1 \\
\vdots & \vdots & \ddots & \vdots \\
t_m^{n-1} & t_m^{n-2} & \cdots & t_m & 1
\end{bmatrix}
\in \mathbb{R}^{m \times n},
\tag{3.16}
$$

and vectors

$$
x \triangleq
\begin{bmatrix}
a_{n-1} \\
a_{n-2} \\
\vdots \\
a_1 \\
a_0
\end{bmatrix}
\in \mathbb{R}^n, \quad
y \triangleq
\begin{bmatrix}
y_1 \\
y_2 \\
\vdots \\
y_m
\end{bmatrix}
\in \mathbb{R}^m.
\tag{3.17}
$$

Then the system of linear equations (3.15) can be represented in a matrix form: $\Phi x = y$. The matrix Φ is known as a *Vandermonde matrix*, and if $m = n$, then Φ is a square matrix and its determinant is given by

$$
\det(\Phi) = \prod_{1 \leq i < j \leq m} (t_i - t_j) = (t_1 - t_2)(t_1 - t_3) \cdots (t_{m-1} - t_m).
\tag{3.18}
$$

It follows that if

$$
t_i \neq t_j, \quad \forall i, j \text{ s.t. } i \neq j,
\tag{3.19}
$$

then Φ is nonsingular and has its inverse. Hence, the solution x^* of (3.15) is given by using Φ^{-1} as

$$
x^* = \Phi^{-1} y.
\tag{3.20}
$$

In summary, if one choose an $(m - 1)$-th order polynomial for m data points that satisfy (3.19), then the coefficients of the interpolating polynomial can be uniquely obtained by the formula (3.20).

Example 3.3. Let us consider the following data.

t	1	2	3	...	14	15
y	2	4	6	...	28	30

The data are obtained from a linear relation $y = 2t$. By using these 15 data points, we find a 14th-order interpolating polynomial. Now, we use a useful computational software, MATLAB, to compute the matrix inversion in (3.20). The following is a code to obtain the coefficients.

MATLAB code for the coefficients of the interpolating polynomial.

```
%% Data
t = 1:15;
y = 2 * t;
%% Vandermonde matrix
Phi = vander(t);
%% Coefficients of interpolating polynomial
x = inv(Phi) * y';
```

In this code, vander is a MATLAB function to compute the Vandermonde matrix in (3.16). The vector variable y computed in the 3rd line is a row vector, and we should transpose it in the last line (i.e. y').

Running this code we obtain

```
x =
    2.274746684520826e-24
   -5.565271161256770e-21
    9.137367918505765e-19
   -9.452887691357992e-18
   -3.658098129966092e-16
   -1.608088662230500e-15
    3.569367024169878e-14
   -6.021849685566849e-13
    5.346834086594754e-13
   -1.267963511963899e-11
    4.878586423728848e-11
    2.088995643134695e-12
    1.366515789413825e-10
    1.999999999995282e+00
   -4.014566457044566e-12
```

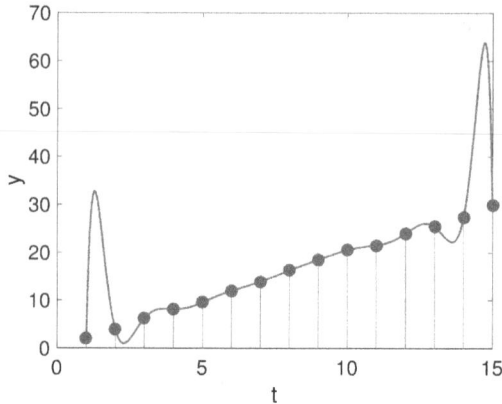

Figure 3.2. 14-th order interpolating polynomial with noisy data.

This result shows that the second value from the bottom is nearly 2, and the other values are almost zero. That is, the coefficients are given by $a_1 = 2$ and $a_i = 0$ for $i \neq 1$, and the interpolating polynomial is $y = 2t$. This is the right solution.   □

Usually, data include *noise*. Let us add Gaussian noise with zero mean and variance 0.5^2 to the data y in Example 3.3, and find the interpolating polynomial. The obtained curve is shown in Figure 3.2. The interpolating polynomial exactly goes through the data points, but the curve is significantly affected by noise, and very different from the original relationship $y = 2t$. Such a phenomenon is called *overfitting*.

The reason of overfitting is that the order of the polynomial is too high. If we previously know that the original curve is of first order, then we can assume a first order polynomial (i.e. a line) $y = a_1 t + a_0$, and find the coefficients a_0 and a_1 with which the line has the best fit to the data. If the data is noisy, it is obviously impossible to obtain a line that goes through all the data points. However, it is not a problem at all if the line does not interpolate the noisy data.

Now, let us reformulate our problem of curve fitting for noisy data. We measure the distance between the polynomial and the data points by the ℓ^2 norm (Euclidean norm). For noisy data, we do not require the curve to go through the data points since it is in general impossible. We find the curve that is as close to the data points as possible. The optimization problem is described as follows:

$$\underset{x \in \mathbb{R}^n}{\text{minimize}} \ \frac{1}{2}\|\Phi x - y\|_2^2, \tag{3.21}$$

where $\Phi \in \mathbb{R}^{m \times n}$ is a Vandermonde matrix defined in (3.16). We call the optimization in (3.21) the *least squares method*. If we assume $n < m$, that is, if the order of the polynomial is less than $m - 2$, then the number of unknowns is less than that of equations. In this case, the matrix Φ is a *tall matrix*, and the equation $\Phi x = y$ has no solution in general. If the condition (3.19) holds, it is easily shown that the solution of (3.21) uniquely exists. In fact, if (3.19) holds, then the Vandermonde matrix Φ has *full column rank*. Note that a matrix $\Phi \in \mathbb{R}^{m \times n}$ is said to have full column rank if the n column vectors in Φ are linearly independent. In other words, $\Phi \in \mathbb{R}^{m \times n}$ has full column rank iff Φ is *injective*, or

$$\text{rank}(\Phi) = n. \tag{3.22}$$

Then the unique solution of the optimization problem in (3.21) is given by

$$x^* = (\Phi^\top \Phi)^{-1} \Phi^\top y. \tag{3.23}$$

We call this the *least squares solution*. As the minimum ℓ^2-norm solution in (3.11), the least squares solution is also given by a closed form.

Exercise 3.4. Prove that the solution of the optimization problem (3.21) is given by (3.23).

Exercise 3.5. Let ϕ_i denote the i-th column vector in matrix $\Phi \in \mathbb{R}^{m \times n}$, that is,

$$\Phi = \begin{bmatrix} \phi_1 & \phi_2 & \cdots & \phi_n \end{bmatrix}. \tag{3.24}$$

Then define the *residual* between the data y and the optimal estimation Φx^* with (3.23) by

$$r \triangleq y - \Phi x^*. \tag{3.25}$$

Prove that the residual satisfies

$$\langle \phi_i, r \rangle = 0, \quad \forall i \in \{1, 2, \ldots, n\}. \tag{3.26}$$

Also, by using this fact, show that the residual r is orthogonal to Φx^*.

Example 3.6. Let us consider Example 3.3 with additive Gaussian noise with zero mean and variance 0.5^2. We assume the curve is a first-order polynomial modeled by $y = a_1 t + a_0$. A MATLAB code to obtain the least squares solution of this problem is given as follows:

```
MATLAB code for least squares solution

%% Data
t = 1:15;
y = 2 * t + randn(1,15)*0.5;
%% Vandermonde matrix
Phi15 = vander(t);
Phi = Phi15(:,14:15);
%% Least squares solution
x = inv(Phi' * Phi) * Phi' * y';
```

In this code, randn(1,15) is a MATLAB function that returns normally distributed random numbers (i.e. Gaussian noise) with zero mean and variance 1 of size 1×15 (i.e. a row vector). The matrix variable Phi15 is a Vandermonde matrix of size 15×15, and in the 6th line we extract 14th and 15th columns, which are related to coefficients a_1 and a_0, to make matrix Phi of size 15×2. The result is shown below.

```
x =
    1.985404378030957e+00
    1.359049380398556e-01
```

Figure 3.3 shows the line $y = a_1 t + a_0$ with these coefficients. While the 14-th order interpolating polynomial implies overfitting, the least squares line shows a good result.   □

3.1.3   Regularization

As we have discussed in the previous section, one can avoid overfitting by the least squares method with an appropriate order of the polynomial, which is less than the number of data points. However, what can we do if we do not know the proper order in advance? In this case, we can adopt *regularization*. Let us begin with a simple example.

Example 3.7. Suppose that we are given a data set

$$\mathcal{D} = \{(t_1, y_1), (t_2, y_2), \ldots, (t_m, y_m)\}, \tag{3.27}$$

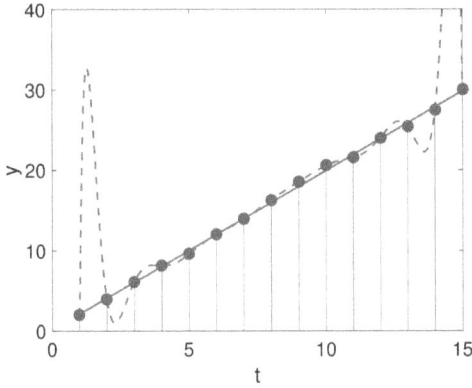

Figure 3.3. Least square solution (solid line) and 14-th order interpolating polynomial (dashed curve).

which is generated from a sinusoid $y = \sin(t)$. We consider sampling instants

$$t_1 = 0, \ t_2 = 1, \ t_3 = 2, \ldots, t_{11} = 10, \tag{3.28}$$

and the points $y_1, y_2, \ldots, y_{11}$ are obtained as

$$y_i = \sin(t_i) + \epsilon_i, \quad i = 1, 2, \ldots, 11, \tag{3.29}$$

where ϵ_i is Gaussian noise with zero mean and variance 0.2^2 added at time t_i independently. The following table shows the obtained data.

t_i	0	1	2	3	4	5
y_i	−0.0343	1.0081	0.8326	0.4047	−0.7585	−0.9285
t_i	6	7	8	9	10	
y_i	−0.2110	0.6626	0.8492	0.2761	−0.6962	

Figure 3.4 shows the data points and the original sinusoidal curve.

For these data, let us find a 10-th order polynomial that interpolates the data points by using (3.20). Figure 3.5 shows the result. Affected by the noise, the curve is very oscillative and shows overfitting. We then take a 6-the order polynomial and compute the least squares solution by (3.23). Figure 3.6 shows the result. From this figure, we have a better fit than the interpolating function shown in Figure 3.5.   □

In the above example, the order 6 was chosen by computing curves of all orders from 1 to 10, and comparing the reconstructed curve with the original sinusoid. However, this can be done *if we previously know the original sinusoid.* This is impossible in real applications. That is, we do not know the optimal order just from the data in advance.

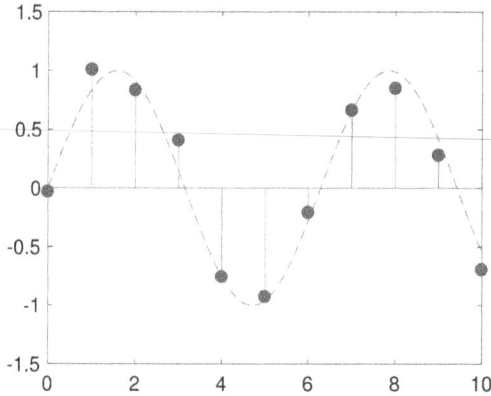

Figure 3.4. 11 data points from a sinusoid (dashed curve).

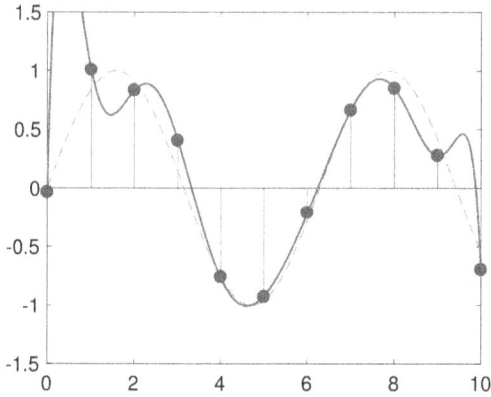

Figure 3.5. 10-th order interpolating polynomial (solid curve) and the original sinusoid (dashed curve).

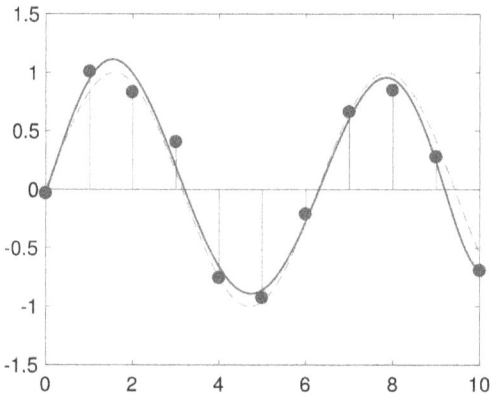

Figure 3.6. Least squares solution with 6-th order polynomial (solid curve) and the original sinusoid (dashed curve).

To see the difference between the 10-th order interpolating polynomial and the 6-th order least squares polynomial, we compare their coefficients. Let us denote by x_{10} and x_6, respectively, the 10-th and 6-th order polynomials. They are obtained as

$$
x_{10} = \begin{bmatrix} -0.0343 \\ 16.2400 \\ \mathbf{-38.0984} \\ \mathbf{37.8369} \\ -20.2842 \\ 6.5035 \\ -1.3100 \\ 0.1677 \\ -0.0133 \\ 0.0006 \\ -0.0000 \end{bmatrix}, \quad x_6 = \begin{bmatrix} -0.0260 \\ \mathbf{1.0636} \\ 0.3067 \\ \mathbf{-0.5225} \\ 0.1426 \\ -0.0146 \\ 0.0005 \end{bmatrix}, \tag{3.30}
$$

where the boldface numbers are the largest three elements in their absolute values. We can observe that the boldface values in x_{10} are much larger than those in x_6. This is a cause of oscillation in the 10-th order interpolating curve.

From the above observation, we try to minimize both the squared error $\|\Phi x - y\|_2^2$ and the squared ℓ^2 norm $\|x\|_2^2$ of the coefficient vector x at the same time. This is formulated as the following optimization problem:

$$
\underset{x \in \mathbb{R}^n}{\text{minimize}} \ \frac{1}{2} \|\Phi x - y\|_2^2 + \frac{\lambda}{2} \|x\|_2^2. \tag{3.31}
$$

We call this optimization the *regularized least squares*, or *ridge regression*. The additional term $\frac{\lambda}{2}\|x\|_2^2$ is called a *regularization term*, and the parameter λ a *regularization parameter*, which control the balance between the error in curve fitting and the ℓ^2 norm of the coefficients.

As in the least squares solution, the solution of the regularized least squares in (3.31) can be obtained in a closed form:

$$
x^* = (\lambda I + \Phi^\top \Phi)^{-1} \Phi^\top y. \tag{3.32}
$$

Exercise 3.8. Prove that the solution of (3.31) is given by (3.32).

Example 3.9. Here we consider an example of regularization. With the data given in Example 3.7, we compute a 10-th order polynomial by the regularized least squares. We take the regularization parameter $\lambda = 1$, and compute the solution x^*

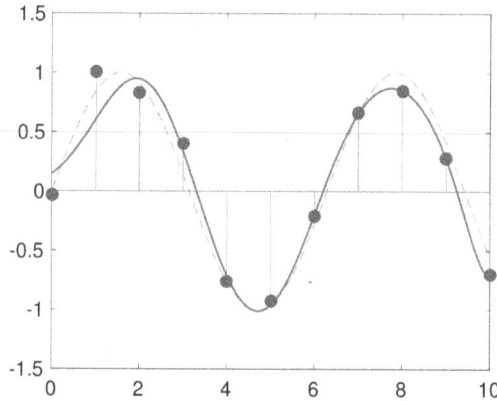

Figure 3.7. Regularized least squares solution with 10th-order polynomial (solid curve) and the original sinusoid (dashed curve).

by the formula (3.32). The obtained coefficients are as follows:

$$
x^* = \begin{bmatrix}
\mathbf{0.1448} \\
\mathbf{0.2691} \\
\mathbf{0.1865} \\
0.0769 \\
-0.0334 \\
-0.0674 \\
0.0386 \\
-0.0085 \\
0.0010 \\
-0.0001 \\
0.0000
\end{bmatrix}.
\tag{3.33}
$$

The boldface values are the three largest elements in the absolute values. Compared with the coefficients x_{10} in (3.30) of 10-th order interpolating polynomial, the values in x^* are much smaller. The curve with the regularized least squares is shown in Figure 3.7. We can see that the 10-th order polynomial by the regularized least squares shows a comparable accuracy to the 6-th order least squares solution. □

3.1.4 Weighted Ridge Regression

Here we further consider the problem of polynomial interpolation. In the regularized least squares, we minimize the cost function in (3.31) with the regularization term $\|x\|_2^2$. This is to make the coefficient vector x not so large. Instead of this, we consider the L^2 norm of the polynomial $f(t)$. The L^2 norm of a function $f(t)$,

$t \in [t_1, t_m]$ is defined as

$$\| f \|_{L^2} \triangleq \sqrt{\int_{t_1}^{t_m} |f(t)|^2 \, dt}. \tag{3.34}$$

Since $f(t)$ is a polynomial

$$f(t) = \sum_{i=0}^{n-1} a_i t^i, \tag{3.35}$$

the L^2 norm can be computed as

$$
\begin{aligned}
\| f \|_{L^2}^2 &= \int_{t_1}^{t_m} \left(\sum_{i=0}^{n-1} a_i t^i \right) \left(\sum_{j=0}^{n-1} a_j t^j \right) dt \\
&= \sum_{i=0}^{n-1} \sum_{j=0}^{n-1} a_i a_j \int_{t_1}^{t_m} t^{i+j} dt \\
&= \sum_{i=0}^{n-1} \sum_{j=0}^{n-1} a_i a_j \frac{t_m^{i+j+1} - t_1^{i+j+1}}{i+j+1} \\
&= x^\top Q x,
\end{aligned}
\tag{3.36}
$$

where $Q = [Q_{ij}]$ is a matrix defined by

$$Q_{ij} = \frac{t_m^{i+j+1} - t_1^{i+j+1}}{i+j+1}, \quad i, j = 0, 1, \ldots, n-1. \tag{3.37}$$

Now, from the definition of L^2 norm, we have $\| f \|_{L^2} \geq 0$ for any polynomial f, and $\| f \|_{L^2} = 0$ if and only if $f = 0$. This means that for any $x \in \mathbb{R}^n$, we have $x^\top Q x \geq 0$ and $x^\top Q x = 0$ if and only if $x = 0$. That is, the matrix Q is *positive definite*.

Now, we consider a regularization problem minimizing

$$\frac{1}{2} \| \Phi x - y \|_2^2 + \frac{\lambda}{2} \| f \|_{L^2}^2 = \frac{1}{2} \| \Phi x - y \|_2^2 + \frac{\lambda}{2} \| \Psi x \|_2^2, \tag{3.38}$$

where Ψ is a matrix that satisfies $Q = \Psi^\top \Psi$. This is called the *weighted ridge regression*. The solution is obtained by

$$x^* = (\Phi^\top \Phi + \lambda \Psi^\top \Psi)^{-1} \Phi^\top y. \tag{3.39}$$

Table 3.1. Summary of optimization problems with ℓ^2 norm.

Problem	Size	Problem	Solution
min ℓ^2 norm	$m < n$	$\min_x \frac{1}{2}\|x\|_2^2$ s.t. $\Phi x = y$	$\Phi^\top(\Phi\Phi^\top)^{-1}y$
least squares (LS)	$m > n$	$\min_x \frac{1}{2}\|\Phi x - y\|_2^2$	$(\Phi^\top\Phi)^{-1}\Phi^\top y$
regularized LS	any	$\min_x \frac{1}{2}\|\Phi x - y\|_2^2 + \frac{\lambda}{2}\|x\|_2^2$	$\Phi^\top(\lambda I + \Phi\Phi^\top)^{-1}y$
			$= (\lambda I + \Phi^\top\Phi)^{-1}\Phi^\top y$

Exercise 3.10. Prove that (3.39) is the solution of the optimization problem minimizing (3.38).

3.1.5 Summary of ℓ^2-Norm Optimization

Now we summarize the curve fitting problem by $(m - 1)$-th order polynomial

$$y = f(t) = a_{n-1}t^{n-1} + a_{n-2}t^{n-2} + \cdots + a_1 t + a_0, \qquad (3.40)$$

with data

$$\mathcal{D} = \{(t_1, y_1), (t_2, y_2), \ldots, (t_m, y_m)\}. \qquad (3.41)$$

Table 3.1 shows the summary.

Exercise 3.11. Prove that for any matrix $\Phi \in \mathbb{R}^{m \times n}$ and any number $\lambda > 0$, matrices $\lambda I + \Phi\Phi^\top$ and $\lambda I + \Phi^\top\Phi$ are invertible and satisfy

$$\Phi^\top(\lambda I + \Phi\Phi^\top)^{-1} = (\lambda I + \Phi^\top\Phi)^{-1}\Phi^\top. \qquad (3.42)$$

3.2 Sparse Polynomial and ℓ^1-norm Optimization

Here we consider yet another example of curve fitting. Let us consider an 80-th order polynomial

$$y = -t^{80} + t. \qquad (3.43)$$

From this polynomial, we sample data points with sampling instants

$$t_1 = 0, t_2 = 0.1, t_3 = 0.2, \ldots, t_{11} = 1, \qquad (3.44)$$

to obtain

$$\mathcal{D} = \{(t_1, y_1), (t_2, y_2), \ldots, (t_{11}, y_{11})\}, \quad y_i = -t_i^{80} + t_i. \qquad (3.45)$$

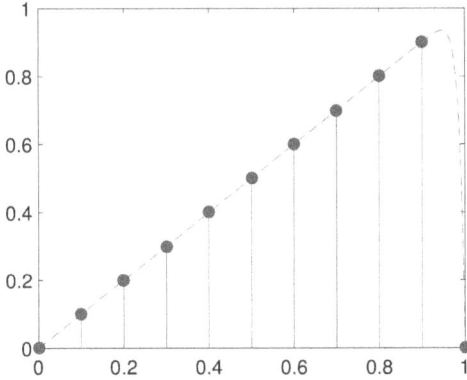

Figure 3.8. Sparse polynomial $y = -t^{80} + t$ and sampled data.

Figure 3.8 shows the curve of the 80-th polynomial in (3.43) and the generated data in (3.45).

We assume that the order of the original polynomial is previously known to be at most 80. Then, can we reconstruct the original curve in (3.43) from the data $\mathcal{D}$? In this case, there are infinitely many interpolating polynomials with order at most 80 that go through all the data points. In fact, the Vandermonde matrix Φ in (3.16) is a fat matrix of size 11×81, and since the condition (3.19) holds, Φ has full row rank and there exist infinitely many solutions of the linear equation $\Phi x = y$, where x is a column vector consisting of 81 unknown coefficients, and y is a column vector consisting of data $y_1, y_2, \ldots, y_{11}$. As mentioned in Section 3.1, we need additional *proofs* to obtain the unique solution.

Let us look again at the original 80-th order polynomial in (3.43). The coefficients of this polynomial are all zero but two coefficients. In other words, the coefficient vector $x = (a_{80}, a_{79}, \ldots, a_0)$ is *sparse*, that is, x has small ℓ^0 norm. We call such a polynomial a *sparse polynomial*. We assume that the following fact can be additionally used as our *proof*.

$$\boxed{\textit{The original polynomial is sparse.}}$$

Note that we can use the sparsity property of the original polynomial but the number of non-zero coefficients (i.e. $\|x\|_0$) is assumed to be unknown.

Borrowing the idea of the optimization mentioned in Section 2.3, we use the ℓ^0 norm as the cost function, and consider the following optimization problem:

$$\underset{x \in \mathbb{R}^n}{\text{minimize}} \ \|x\|_0 \ \text{ subject to } \ \Phi x = y. \tag{3.46}$$

As mentioned in Section 2.4, this is quite hard to solve using the exhaustive search method when the problem size is large.

The key idea of sparse optimization is to use the ℓ^1 *norm*

$$\|x\|_1 = \sum_{i=1}^{n} |x_i| \qquad (3.47)$$

instead of the ℓ^0 norm. That is, we consider the following optimization problem as relaxation of the ℓ^0 optimization (3.46):

$$\underset{x \in \mathbb{R}^n}{\text{minimize}} \ \|x\|_1 \quad \text{subject to} \ \Phi x = y. \qquad (3.48)$$

We call this optimization the ℓ^1 *optimization*. The method to obtain a sparse vector by the ℓ^1 optimization is known as the *basis pursuit*.

The ℓ^1 optimization problem in (3.48) is to find the smallest ℓ^1-norm vector on a linear subspace $\{x \in \mathbb{R}^n : \Phi x = y\}$. As illustrated in Figure 2.2 in Chapter 2, the contour of the ℓ^1 norm ($\|x\|_1 = c$) is a diamond whose corners are on the axes. The optimal solution of (3.48) is obtained (in the 2-dimensional case) by enlarging the contour $\|x\|_1 = c$ from $c = 0$ until the contour touches the linear subspace $\{x \in \mathbb{R}^2 : \Phi x = y\}$. As shown in Figure 3.9, the linear subspace $\{x \in \mathbb{R}^2 : \Phi x = y\}$ touches almost surely the ℓ^1 contour on one of its corners. Since each corner is on one of the axes, the optimal solution satisfies $\|x^*\|_0 = 1 < 2$, and hence it is sparse. This is an intuitive explanation why minimizing ℓ^1 norm gives a sparse solution.

The relation between the ℓ^0 and ℓ^1 norms is intuitively understood as follows. By definition, the ℓ^0 norm can be rewritten as

$$\|x\|_0 = \sum_{i=1}^{n} |x_i|^0, \qquad (3.49)$$

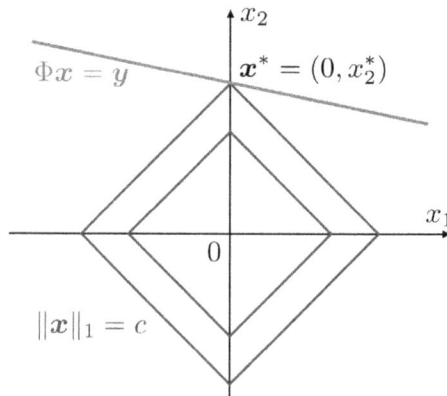

Figure 3.9. ℓ^1 optimization in $\mathbb{R}^2$: the contour $\{x : \|x\|_1 = c\}$ touches the linear subspace $\{x : \Phi x = y\}$ on one of the corners that are on axes.

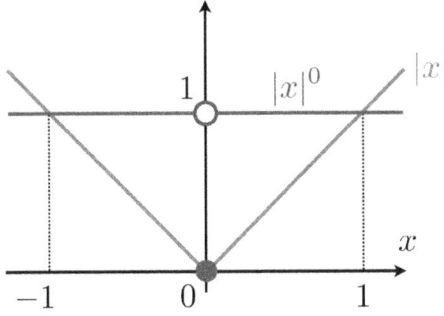

Figure 3.10. Relation between $|x|$ and $|x|^0$.

where $|x|^0 \triangleq 1$ if $x \neq 0$ and $0^0 \triangleq 0$. Figure 3.10 shows the graph of $|x|^0$. From this figure, it is easily seen that the ℓ^0 norm is non-convex. On the other hand, the ℓ^1 norm

$$\|\boldsymbol{x}\|_1 = \sum_{i=1}^{n} |x_i| \tag{3.50}$$

is a sum of absolute values $|x_i|$, which is convex as shown in Figure 3.10. The ℓ^1 norm is the best convex approximation of the ℓ^0 norm in the sense that it has the minimum exponent $p = 1$ among ℓ^p norms that are convex. Theoretically, the ℓ^1 norm is also understood as the *convex relaxation* of the ℓ^0 norm. That is, the ℓ^1 norm is the second conjugate $\| \cdot \|_0^{**}$ of $\| \cdot \|_0$. See [112, Section 1.3] for details.

In the ℓ^1 optimization problem in (3.48), the cost function $\|\boldsymbol{x}\|_1$ is a convex function of $\boldsymbol{x}$, and the constraint set $\{\boldsymbol{x} \in \mathbb{R}^n : \Phi\boldsymbol{x} = \boldsymbol{y}\}$ is a convex set. Therefore, the problem is a *convex optimization problem*,[1] for which *numerical optimization* by using a computer can give a numerical solution much faster than the exhaustive search for the ℓ^0 optimization in (3.46).

To obtain a sparse vector, we can also use the idea of regularization for sparse optimization. In the case of noisy data, we can formulate the curve fitting as the ℓ^0 *regularization* described below:

$$\underset{\boldsymbol{x} \in \mathbb{R}^n}{\text{minimize}} \; \frac{1}{2}\|\Phi\boldsymbol{x} - \boldsymbol{y}\|_2^2 + \lambda\|\boldsymbol{x}\|_0. \tag{3.51}$$

Unfortunately, this optimization is also a combinatorial problem, and hard to solve if the problem size is large. Instead of the ℓ^0 norm for the regularization term, we use the ℓ^1 norm and consider the following optimization problem:

$$\underset{\boldsymbol{x} \in \mathbb{R}^n}{\text{minimize}} \; \frac{1}{2}\|\Phi\boldsymbol{x} - \boldsymbol{y}\|_2^2 + \lambda\|\boldsymbol{x}\|_1 \tag{3.52}$$

---

1. For the mathematical definition of convexity and convex optimization, see Chapter 4.

This is called ℓ^1 *regularization*, or *LASSO* (Least Absolute Shrinkage and Selection Operator). The cost function in (3.52) is a convex function of x, and hence the optimization is a convex optimization problem, which can also be solved very efficiently.

In this section, we have introduced the important idea to approximate the ℓ^0 norm, which is non-convex and discontinuous, by the ℓ^1 norm, which is convex. A question is when the convex optimization problem (3.48), or (3.52) can give the solution of the original ℓ^0 optimization (3.46), or (3.51). Very interestingly, in many applications (e.g. signal/image processing), the solution of the ℓ^1 norm optimization is equivalent to (or sufficiently close to) the ℓ^0-norm solution. In fact, there exist many theorems for the equivalence between ℓ^0 and ℓ^1 optimizations. From these facts, ℓ^1 optimization is often said to be sparse optimization. For seeking sparse solutions, there also exist many methods other than ℓ^1 optimization, for example, greedy methods, which we will see in detail in Chapter 5, or ℓ^p-norm optimization with $p \in (0, 1)$. In the next section, we will show how to solve the ℓ^1 optimization (3.48) or ℓ^1 regularization (3.52) by using MATLAB.

3.3   Numerical Optimization by CVX

The optimization problems (3.48) and (3.52) are convex ones, and they can be efficiently solved by numerical optimization. We here introduce a well-known software for numerical convex optimization, CVX,[2] which is a free software running on MATLAB.

Let us consider the 80-th order polynomial $y = -t^{80} + t$ in (3.43). From this, we generate 11 data points as in (3.45), and we try to reconstruct the original polynomial from these data.

First, we define the coefficient vector of the 80-th polynomial. The following is a MATLAB code to do this.

```
%% Coefficient vector of the 80-th order polynomial
x_orig = [-1,zeros(1,78),1,0]';
```

---

2.   http://cvxr.com/cvx/

Next, by using MATLAB function polyval that returns the value of the polynomial from the coefficient vector, we make a data set (3.45) as follows.

```
%% Data
t = 0:0.1:1;
y = polyval(x_orig,t);
```

With the data, we find the 10-th order interpolating polynomial. For this, we compute the Vandermonde matrix (3.16):

```
%% Vandermonde matrix
Phi = vander(t);
```

From this, we compute the coefficients of the interpolating polynomial by using (3.20).

```
%% Coefficients of interpolating polynomial (10-th order)
x = inv(Phi) * y';
```

Note that since y is a row vector, we take its transpose y'. Now, let us draw the curve of the interpolating polynomial. We discretize the time axis into small intervals, and draw the curve.

```
%% Draw curve
time = 0:0.01:1;
plot(time, polyval(x, time));
```

Figure 3.11 shows the result. In this case, the data are noiseless and there is no oscillation due to overfitting. However, we can see a large gap in the range from $t = 0.9$ to $t = 1$.

Then, let us compute a curve by the regularized least squares (3.31) with a 10-th order polynomial. We choose the regularization parameter as $\lambda = 0.2$, and compute the coefficient by the formula (3.32).

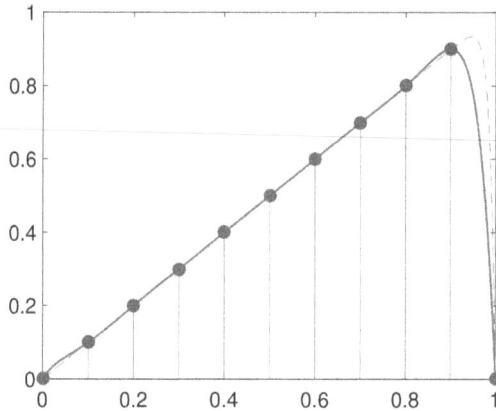

Figure 3.11. 10-th order interpolating polynomial (solid curve) and the original polynomial $y = -t^{80} + t$ (dashed curve).

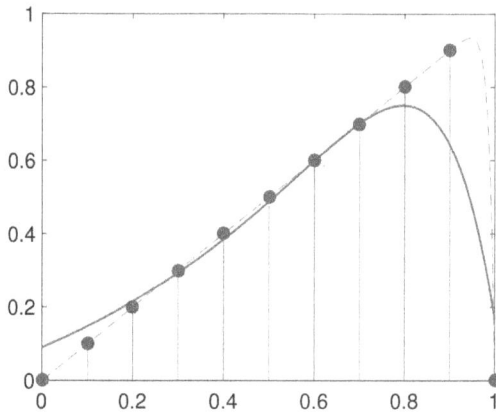

Figure 3.12. 10-th order polynomial by regularized least squares (solid curve) and the original polynomial $y = -t^{80} + t$ (dashed curve).

```
%% Regularized least squares
lambda = 0.2;
x = inv(lambda * eye(11) + Phi' * Phi) * Phi' * y';
```

Figure 3.12 shows the result. The obtained curve has a poor fit to the original curve $y = -t^{80} + t$.

Finally, we compute the curve by ℓ^1 optimization (3.48). We assume that we know the polynomial order is at most 80. In this case, the Vandermonde matrix (3.16) becomes a fat matrix of size 11×81. This matrix can be obtained as follows.

```
%% Vandermode's matrix
Phi = [ ];
for m = 0:80
  Phi = [t'.^m, Phi];
end
```

Define the coefficient vector x and the data vector y as in (3.17), then the condition for interpolation is described as

$$\Phi x = y. \tag{3.53}$$

We seek the sparsest solution in the linear subspace $\{x \in \mathbb{R}^{81} : \Phi x = y\}$ by solving the ℓ^1 optimization problem in (3.48).

To solve the ℓ^1 optimization problem, we use CVX on MATLAB. By using CVX, the optimization problem can be very easily coded.

```
%% L1 optimization by CVX
cvx_begin
  variable x(81)
  minimize norm(x, 1)
  subject to
    Phi * x == y'
cvx_end
```

You should compare this code with (3.48). You can write a code very intuitively for an optimization problem. This is the strongest point of CVX. You can solve many convex optimization problems other than the ℓ^1 optimization in a similar way. We recommend for beginners to use CVX to solve convex optimization problems.[3]

Let us draw the curve with the coefficients obtained by the ℓ^1 optimization. Figure 3.13 shows the curve. We can see that the obtained curve is almost the same as the original curve $y = -t^{80} + t$. This is the power of ℓ^1 optimization. The complete MATLAB program for ℓ^1 optimization is shown below. Enjoy!

---

3.   CVX is also available in Python. See cvxopt.org for details.

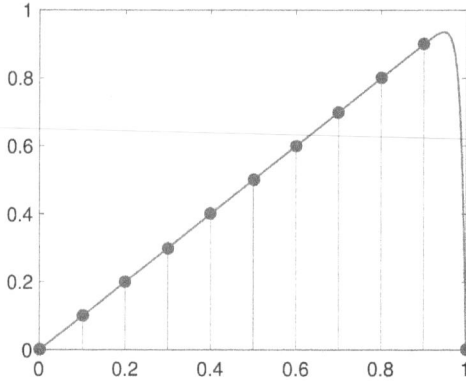

Figure 3.13. 80-th order polynomial by ℓ^1 optimization (solid curve) and the original polynomial $y = -t^{80} + t$ (dashed curve).

3.4   Further Readings

For least squares, regularization and overfitting in curve fitting, I recommend to read standard textbooks [8, 44] of machine learning. The least squares method is deeply related to projection and generalized inverse, for which you can choose a textbook of [43]. The mathematical theory and generalization of LASSO can be found in [15, 40, 44, 45]. For the equivalence theorems between ℓ^0 and ℓ^1 optimizations, refer to text books [37, 38, 112].

MATLAB program for the coefficients by ℓ^1 optimization using CVX.

```
clear
%% Polynomial coefficients
x_orig = [-1,zeros(1,78),1,0]';
%% Sampling
t = 0:0.1:1;
y = polyval(x_orig,t);
%% Data size
N = length(t);
M = N-1;
%% Vandermonde matrix
Phi_v = vander(t);
%% Interpolation polynomial with order 10
x_i = inv(Phi_v)*y';
%% LASSO
```

```
% Order of polynomial
M_l = 80;
% Design matrix
Phi_l=[];
for m=0:M_l
 Phi_l = [t'.^m,Phi_l];
end
% CVX
cvx_begin
 variable x(M_l+1)
 minimize norm(x,1)
 subject to
   Phi_l*x == y'
cvx_end
%% Plot
tt = 0:0.01:1;
figure;
stem(t,y); hold on
plot(tt,polyval(x_orig,tt),'-');
plot(tt,polyval(x,tt));
```

DOI: 10.1561/9781680837254.ch4

Chapter 4

Algorithms for Convex Optimization

In the previous chapter, we have seen that convex optimization such as ℓ^1 optimization is efficiently solved by using CVX on MATLAB. Such a tool is actually very useful for small or middle scale problems. However, if you treat a very large-scale problem like image processing, CVX might be insufficient. Moreover, if you want to apply the sparsity method to control systems, you should compute sparse optimization in real time (e.g. in a few msec) with a cheap device on which MATLAB cannot be installed. In such cases, you should instead write an efficient algorithm by yourself for your specific problem. This means that you should look into the *black box* of the toolbox.

For this purpose, we review basics of convex optimization, and introduce efficient algorithms for problems in sparse optimization.

Key ideas of Chapter 4

- In convex optimization, a local minimum is a global minimum.
- ℓ^1 optimization problems appeared in this book are convex optimization.
- Proximal operators are used to derive fast algorithms for convex optimization with non-differentiable ℓ^1 norm and constraints.

4.1 Basics of Convex Optimization

We here review important facts in convex optimization. Let us begin with the definition of a convex set.

Definition 4.1 (convex set). *Let C be a subset of $\mathbb{R}^n$. C is said to be a* convex set *if the following inclusion*

$$tx + (1 - t)y \in C \tag{4.1}$$

holds for any vectors $x, y \in C$ and for any real number $t \in [0, 1]$.

Figure 4.1 illustrates a convex set and a *non-convex set* (i.e. a set that is not convex). In convex set C, the line segment between any two points x and y in C lies completely in C. On the other hand, in a non-convex set, there exists a line segment that partially lies outside of the set.

Exercise 4.2. Prove that if C and $\mathcal{D}$ are convex, then $C \cap \mathcal{D}$ is also convex.

In convex optimization, we often handle a function $f : \mathbb{R}^n \to \mathbb{R} \cup \{\infty\}$, which takes values on *extended real numbers* $\mathbb{R} \cup \{\infty\}$. The following function is an example:

$$f(x) = \begin{cases} 0, & \text{if } \|x\|_2 \le 1, \\ \infty, & \text{if } \|x\|_2 > 1. \end{cases} \tag{4.2}$$

This function is called an indicator function, which will be explained in Section 4.2.4.

The *effective domain* of a function f is defined by

$$\text{dom}(f) \triangleq \{x \in \mathbb{R}^n : f(x) < \infty\}. \tag{4.3}$$

That is, the effective domain of a function $f : \mathbb{R}^n \to \mathbb{R} \cup \{\infty\}$ is a set in $\mathbb{R}^n$ on which f takes finite real values. For example, the effective domain of the indicator function (4.2) is given by

$$\text{dom}(f) = \{x \in \mathbb{R}^n : \|x\|_2 \le 1\}. \tag{4.4}$$

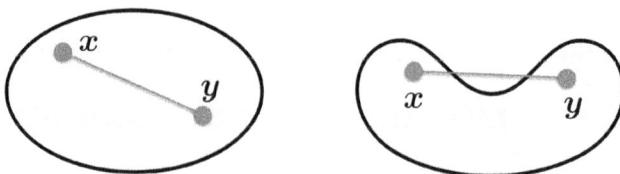

Figure 4.1. Convex set (left) and non-convex set (right).

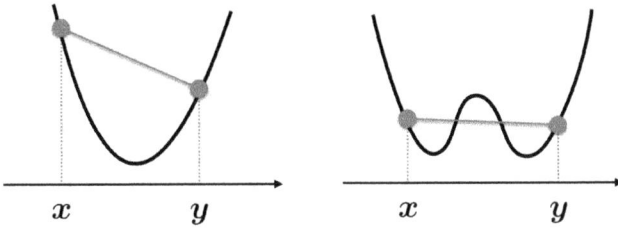

Figure 4.2. Convex function (left) and non-convex function (right).

A function is said to be *proper* if its effective domain is non-empty, that is, there exists at least one $x \in \mathbb{R}^n$ such that $f(x) < \infty$.

Now, let us define a convex function.

Definition 4.3 (convex function). *Let $f : \mathbb{R}^n \to \mathbb{R} \cup \{\infty\}$ be a proper function. The function f is said to be a* convex function *if the following inequality*

$$f(tx + (1-t)y) \leq tf(x) + (1-t)f(y) \tag{4.5}$$

holds for any vectors $x, y \in \text{dom}(f)$ and for any real number $t \in [0, 1]$.

Figure 4.2 illustrates a convex function and a *non-convex function*, a function that is not convex. By definition, if f is convex, the line segment between any two points $(x, f(x))$ and $(y, f(y))$, where $x, y \in \text{dom}(f)$, lies above or on the graph of f. On the other hand, if f is non-convex, there exists a line segment that partially lies below the graph.

Exercise 4.4. Suppose that f and g are real-valued convex functions. Prove that $f + g$ is also convex.

One more important property of a function is *closedness*. A function $f : \mathbb{R} \to \mathbb{R} \cup \{\infty\}$ is said to be *closed* if the *sublevel set* (or *lower level set*) $\{x \in \text{dom}(f) : f(x) \leq c\}$ is a closed set for any $c \in \mathbb{R}$. The closedness of a function is also understood by its *epigraph*. The epigraph epi(f) of function f is defined by

$$\text{epi}(f) \triangleq \{(x, t) \in \mathbb{R}^n \times \mathbb{R} : x \in \text{dom}(f), f(x) \leq t\}. \tag{4.6}$$

Figure 4.3 illustrates the epigraph of a function f. The epigraph of f is the region above the graph on its effective domain. It is easily shown that a function f is closed if and only if its epigraph is closed.

We can also perceive other properties of a function in terms of its epigraph. A function is convex if and only if its epigraph is convex. A function is proper if and only if its epigraph is non-empty. We summarize these facts in Table 4.1.

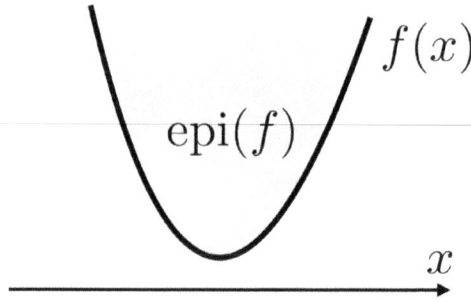

Figure 4.3. Epigraph epi(f) of function f.

Table 4.1. Function and its epigraph.

Function f	Epigraph Epi(f)
convex	convex set
closed	closed set
proper	non-empty set

Now, we formulate a convex optimization problem in a general form.

┌─ Convex optimization problem ─────────────────────────────────┐

Let $f : \mathbb{R}^n \to \mathbb{R} \cup \{\infty\}$ be a proper, closed, and convex function, and $C \subset \mathbb{R}^n$ be a closed convex set. Then, a *convex optimization problem* is a problem to find a vector $x^* \in \mathbb{R}^n$ that minimizes the function $f(x)$ over the set $C \subset \mathbb{R}^n$.

└───┘

For the convex optimization, we use the following terminology in this book:

- The function $f(x)$ is called a *cost function* or an *objective function*.
- The set C is called a *constraint set* or a *feasible set*.
- The entries of C are called *feasible solutions*.
- The inclusion $x \in C$ is called a *constraint*

The above optimization problem is often described as follows:

$$\underset{x \in \mathbb{R}^n}{\text{minimize}} \ f(x) \ \text{subject to} \ x \in C. \qquad (4.7)$$

In this expression, the optimization variable $x \in \mathbb{R}^n$ to be minimized is placed under "minimize," next to which the cost function $f(x)$ is placed. The term "subject to" is sometimes abbreviated as "s.t.," followed by the constraint $x \in C$. The term "minimize" in (4.7) is often abbreviated as "min" and simply described as

$$\underset{x \in \mathbb{R}^n}{\min} \ f(x) \ \text{s.t.} \ x \in C. \qquad (4.8)$$

$$\underset{x \in \mathbb{R}^n}{\text{minimize}} \;\; \underbrace{f(x)}_{\text{cost function}} \;\; \text{subject to} \;\; \underbrace{x \in C}_{\text{constraint}}$$

cost function   constraint

$$\underset{x \in C}{\min} f(x) \qquad \text{minimum value}$$

$$\underset{x \in C}{\arg \min} \; f(x) \qquad \text{minimizer (set)}$$

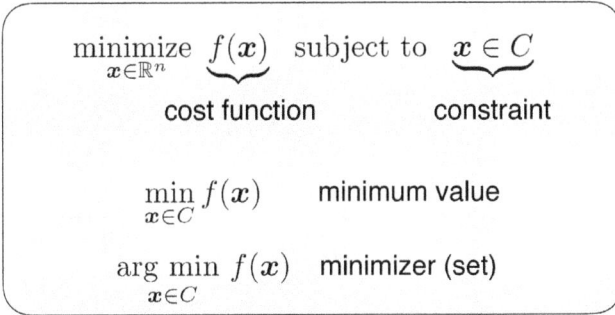

Figure 4.4. Notation for optimization problem.

Also, we often write the constraint under "minimize" as

$$\underset{x \in C}{\min} f(x). \tag{4.9}$$

Note that (4.9) usually means the minimum value of the optimization problem (4.7), instead of an optimization problem. The set of minimizers (solutions) of the optimization problem (4.7) is denoted using "arg" (abbreviation of argument) as

$$\underset{x \in C}{\arg \min} \, f(x) \triangleq \left\{ x^* \in C : f(x^*) \le f(x), \quad \forall x \in C \cap \operatorname{dom}(f) \right\}. \tag{4.10}$$

Also, we often use the following expression

$$x^* = \underset{x \in C}{\arg \min} \, f(x). \tag{4.11}$$

In this expression, "argmin" returns a minimizer, instead of the set of minimizers. If the minimizer of the optimization problem (4.7) is unique, then this expression may not cause any confusion. If not unique, (4.11) means that x^* is a minimizer arbitrarily taken from the set of minimizers.

We summarize the definitions in Figure 4.4.

Then we define a local minimizer and a global minimizer of the optimization problem (4.7). If there exists an open set $\mathcal{B}$ that contains a feasible solution $\bar{x} \in C \cap \operatorname{dom}(f)$ such that

$$f(x) \ge f(\bar{x}), \quad \forall x \in \mathcal{B} \cap C, \tag{4.12}$$

then $f(\bar{x})$ is called a *local minimizer* of the optimization problem (4.7). If a feasible solution $x^* \in C$ satisfies

$$f(x) \ge f(x^*), \quad \forall x \in C, \tag{4.13}$$

then x^* is called a *global minimizer* of the optimization problem (4.7).

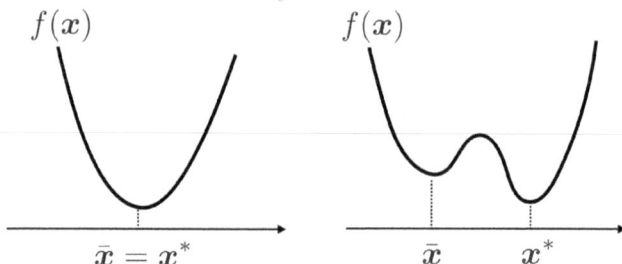

Figure 4.5. Local minimizer $\bar{x}$ and global minimizer x^* with convex function (left) and non-convex function (right).

One of the most important properties of convex optimization is that a local minimizer is (if it exists) a global minimizer. Figure 4.5 illustrates this fact of convex optimization. In this figure, for a convex function, the local minimizer $\bar{x}$ is also the global minimizer x^*. On the other hand, for a non-convex function, they may not coincide. In fact, the following theorem holds [13, Section 4.2.2].

Theorem 4.5. *For a convex optimization problem (4.7), any local minimizer is (if it exists) a global minimizer, and the set of global minimizers is a convex set.*

By this theorem, an algorithm that outputs a local minimizer of a convex optimization problem is automatically an algorithm for a global minimizer. For example, a convex optimization problem with a differentiable and convex function $f(x)$ and $C = \mathbb{R}^n$ (unconstrained problem), a point $\bar{x}$ such that $\nabla f(\bar{x}) = 0$, where ∇f is the *gradient* of f, is a local minimizer, and this is also a global minimizer. Therefore, for an unconstrained convex optimization with a differentiable cost function, an algorithm searching for a point satisfying $\nabla f(x) = 0$ is an algorithm for a global minimizer. This idea is very important to derive an efficient algorithm for convex optimization.

Exercise 4.6. Find a convex function that has no local minimizer.

Exercise 4.7. Find a convex function that has infinitely many local minimizers.

Next, we consider the uniqueness of the minimizer. For this, we define strictly and strongly convex functions.

Definition 4.8. *Let $f : \mathbb{R}^n \to \mathbb{R} \cup \{\infty\}$ be a proper function. The function f is said to be a* strictly convex function *if for any $x, y \in \text{dom}(f) \subset \mathbb{R}^n$ with $x \neq y$ and any $t \in (0, 1)$,*

$$f(tx + (1-t)y) < tf(x) + (1-t)f(y) \tag{4.14}$$

Moreover, the function f is said to be a strongly convex function *if there exists* $\beta > 0$ *such that for any* $x, y \in \mathrm{dom}(f) \subset \mathbb{R}^n$ *and any* $t \in [0, 1]$,

$$f(tx + (1-t)y) \le tf(x) + (1-t)f(y) - t(1-t)\frac{\beta}{2}\|x - y\|_2^2 \quad (4.15)$$

The constant β *is called a* modulus.

The following lemma is an important property of strongly convex functions.

Lemma 4.9. *A function* $f : \mathbb{R}^n \to \mathbb{R} \cup \{\infty\}$ *is strongly convex with modulus* $\beta > 0$ *if and only if*

$$f - \frac{\beta}{2}\| \cdot \|_2^2 \quad (4.16)$$

is convex.

Weierstrass extreme value theorem is also important in convex optimization.

Theorem 4.10 (Weierstrass extreme value theorem). *Every continuous function on a compact set attains its extreme values on that set.*

Note that a subset in $\mathbb{R}^n$ is compact if and only if it is closed and bounded.

The following theorem shows the existence and uniqueness of the minimizer of a strongly convex function.

Theorem 4.11. *Assume* $f : \mathbb{R}^n \to \mathbb{R} \cup \{\infty\}$ *is a proper, closed, and strongly convex function with modulus* $\beta > 0$. *Then* f *has the unique minimizer* $x^* \in \mathrm{dom}(f)$. *That is, for all* $x \in \mathrm{dom}(f)$ *such that* $x \ne x^*$,

$$f(x) > f(x^*). \quad (4.17)$$

Moreover, for any $x \in \mathrm{dom}(f)$, *we have*

$$f(x) \ge f(x^*) + \frac{\beta}{2}\|x - x^*\|_2^2. \quad (4.18)$$

This theorem is used to define the proximal operator discussed in the next section.

Exercise 4.12. Prove Theorem 4.11.

4.2 Proximal Operator

We here introduce a powerful tool called the proximal operator for deriving efficient algorithms to solve convex optimization problems.

4.2.1 Definition

The proximal operator of a function is defined as follows:

Definition 4.13 (proximal operator). *For a proper, closed, and convex function* $f : \mathbb{R}^n \to \mathbb{R} \cup \{\infty\}$, *and a real number* $\gamma > 0$, *the* proximal operator $\text{prox}_{\gamma f}$ *with parameter* γ *is defined by*

$$\text{prox}_{\gamma f}(v) \triangleq \underset{x \in \text{dom}(f)}{\arg\min} \left\{ f(x) + \frac{1}{2\gamma} \|x - v\|_2^2 \right\}. \tag{4.19}$$

First, we can easily show that the function

$$g(x) \triangleq f(x) + \frac{1}{2\gamma} \|x - v\|_2^2 \tag{4.20}$$

is a proper, closed, and strongly convex function with modulus $\beta = 1/\gamma$ (see Definition 4.8). Therefore, from Theorem 4.11, the proximal operator (4.19) is well-defined, that is, $\text{prox}_{\gamma f}(v)$ uniquely exists for any $v \in \mathbb{R}^n$.

Exercise 4.14. Assume that f is a proper, closed, and convex function, $\gamma > 0$, and $v \in \mathbb{R}^n$. Prove that the function $g(x)$ in (4.20) is proper, closed, and strongly convex function with modulus $\beta = 1/\gamma$.

From (4.19), if we take $\gamma \to \infty$, then the second term of (4.19) disappears and the proximal operator becomes

$$\text{prox}_{\infty f}(v) = \underset{x \in \text{dom}(f)}{\arg\min} \ f(x) = x^*, \tag{4.21}$$

where x^* is a minimizer of $f(x)$. On the other hand, taking $\gamma \to 0$ eliminates the first term of (4.19), and the proximal operator is reduced to

$$\text{prox}_{0f}(v) = \underset{x \in \text{dom}(f)}{\arg\min} \ \|x - v\|_2^2 = \Pi_C(v), \quad C \triangleq \text{dom}(f), \tag{4.22}$$

where Π_C is the *projection operator* on the set C. That is, Π_C returns the closest point in C measured by the ℓ^2 norm. Finally, if the parameter γ satisfies $0 < \gamma < \infty$, the proximal operator (4.19) is a mixture of the minimizer in (4.21) and the projection operator in (4.22).

Figure 4.6 illustrates the proximal operator. By definition, if a point v is outside the effective domain $\text{dom}(f)$, then $\text{prox}_{\gamma f}(v)$ moves into $\text{dom}(f)$. If a point v is in $\text{dom}(f)$, then $\text{prox}_{\gamma f}(v)$ moves in $\text{dom}(f)$, and approaches towards the minimizer x^* of $f(x)$, with a step size determined by the value of γ. Therefore, the effective domain $\text{dom}(f)$ is an *invariant set* under the proximal operator $\text{prox}_{\gamma f}$. Note that

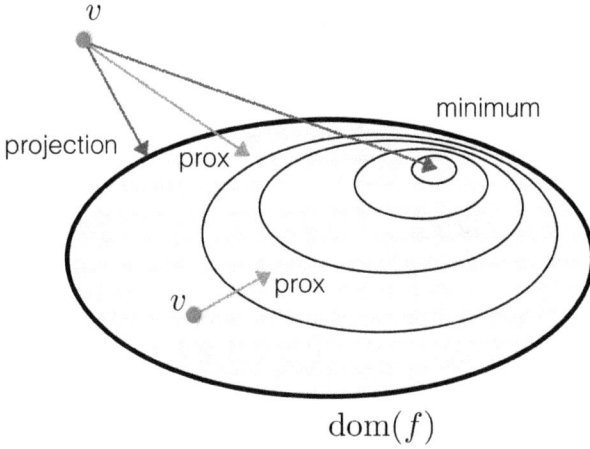

Figure 4.6. Illustration of proximal operator.

a set C is called an invariant set under an operator T if

$$x \in C \implies T(x) \in C \tag{4.23}$$

holds.

Exercise 4.15. Prove that the effective domain $\text{dom}(f)$ is an invariant set under the proximal operator $\text{prox}_{\gamma f}$.

As illustrated in Figure 4.6, if $v \in \text{dom}(f)$, then the vector $\text{prox}_{\gamma f}(v)$ approaches towards the minimizer x^* in the effective domain $\text{dom}(f)$. That is, a proximal operator works like a *negative gradient* in the effective domain. Note that the gradient cannot be defined for non-differentiable functions, while the proximal operator has no such restriction.

4.2.2 Proximal Algorithm

From the invariance property of (4.23), we can consider an iterative algorithm called the *proximal algorithm* for a minimizer x^* of function f:

Proximal algorithm

Initialization: give an initial vector $x[0]$ and positive numbers $\gamma_0, \gamma_1, \ldots$
Iteration: for $k = 0, 1, 2, \ldots$, do

$$x[k + 1] = \text{prox}_{\gamma_k f}(x[k]). \tag{4.24}$$

If you properly choose the parameter sequence $\{\gamma_k\}$, you can obtain one of the minimizers of f by the proximal algorithm. The convergence is shown in the following theorem [7, Proposition 5.1.3]:

Theorem 4.16 (convergence of proximal algorithm). *Suppose that the parameter sequence* $\{\gamma_k\}$ *satisfies* $\gamma_k > 0$ *for all* k *and*

$$\sum_{k=0}^{\infty} \gamma_k = \infty. \tag{4.25}$$

Then, the vector sequence $\{x[k]\}$ *generated by the proximal algorithm* (4.24) *converges to one of the minimizers of* f *for any initial vector* $x[0]$.

The theorem is based on the fact that a minimizer of $f(x)$ is also a *fixed point* of its proximal operator $\text{prox}_{\gamma f}$. Note that a fixed point of $\text{prox}_{\gamma f}$ is a point that satisfies

$$x = \text{prox}_{\gamma f}(x). \tag{4.26}$$

A fixed point is literally *fixed* under the operation by $\text{prox}_{\gamma f}$.

The proximal algorithm minimizes the *strongly convex* function

$$g_k(x) \triangleq f(x) + \frac{1}{2\gamma_k} \|x - x[k]\|_2^2 \tag{4.27}$$

at step k. In other words, the algorithm approximates a general convex function $f(x)$ by a strongly convex function at each step.

Also, it is often important to find a closed form of the proximal operator for an efficient algorithm. A function for which the proximal operator (4.19) is obtained in a closed form is sometimes called *proximable*. Let us see some proximable functions in the following subsections.

4.2.3   Proximal Operator for Quadratic Function

Let us consider the following *quadratic function*

$$f(x) = \frac{1}{2} x^{\top} \Phi x - y^{\top} x, \tag{4.28}$$

where Φ is a *positive-definite* symmetric matrix. Note that a symmetric matrix Φ is said to be *positive definite* if the following inequality holds

$$x^{\top} \Phi x > 0, \tag{4.29}$$

for every nonzero vector $x \in \mathbb{R}^n$. Let us compute the proximal operator of the quadratic function in (4.28). From the definition of the proximal operator in (4.19), we have

$$\text{prox}_{\gamma f}(v) = \arg\min_{x \in \mathbb{R}^n} \left\{ \frac{1}{2} x^{\top} \Phi x - y^{\top} x + \frac{1}{2\gamma} (x - v)^{\top} (x - v) \right\}. \tag{4.30}$$

Since the function in (4.30) is differentiable, we can obtain the minimizer by setting the gradient to be zero. After some calculations, we have the proximal operator in a closed form:

$$\text{prox}_{\gamma f}(v) = \left(\Phi + \frac{1}{\gamma}I\right)^{-1}\left(y + \frac{1}{\gamma}v\right). \tag{4.31}$$

Exercise 4.17. Prove that equation (4.31) holds.

An important application of this proximal operator is numerical matrix inversion. The minimizer x^* of (4.28) is also the unique solution of linear equation

$$\Phi x = y, \tag{4.32}$$

that is, $x^* = \Phi^{-1}y$. Note that Φ is invertible since Φ is positive definite. Then, let us assume that the *condition number* (the ratio of maximum and minimum eigenvalues of Φ, i.e., $\lambda_{\max}(\Phi)/\lambda_{\min}(\Phi)$) is very large so that the numerical computation of the inverse is difficult. We call such a case *ill-conditioned*. For an ill-conditioned case, the proximal algorithm (4.24) is used to safely compute the inverse. From (4.31), the proximal algorithm to obtain the minimizer of (4.28), which is also the solution of (4.32), is given as follows:

Proximal algorithm for $\Phi^{-1}y$

Initialization: give an initial vector $x[0]$ and a positive number $\gamma > 0$
Iteration: for $k = 0, 1, 2, \ldots$, do

$$x[k+1] = \left(\Phi + \frac{1}{\gamma}I\right)^{-1}\left(y + \frac{1}{\gamma}x[k]\right). \tag{4.33}$$

If the positive number γ is sufficiently small, then the condition number of matrix $\Phi + (1/\gamma)I$ is relatively small, and the inversion can be easily computed numerically.

Also, if γ is sufficiently small, we have

$$\left(\Phi + \frac{1}{\gamma}I\right)^{-1} = \gamma(I + \gamma\Phi)^{-1} \approx \gamma(I - \gamma\Phi). \tag{4.34}$$

Then, the right-hand side of the proximal operator (4.31) becomes

$$\left(\Phi + \frac{1}{\gamma}I\right)^{-1}\left(y + \frac{1}{\gamma}v\right) \approx \gamma(I - \gamma\Phi)\left(y + \frac{1}{\gamma}v\right)$$

$$\approx v - \gamma(\Phi v - y) \tag{4.35}$$

$$= v - \gamma\nabla f(v).$$

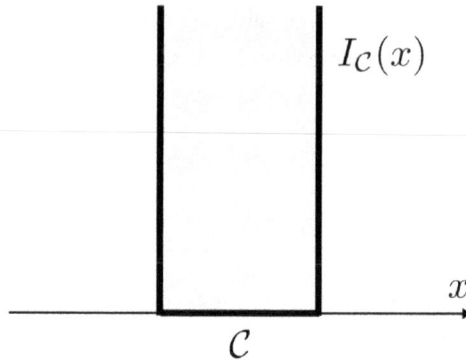

Figure 4.7. Indicator function $I_C(x)$ on a closed interval $C \in \mathbb{R}$.

That is, if γ is sufficiently small, the proximal algorithm (4.33) is approximately the *gradient descent algorithm*

$$x[k + 1] = x[k] - \gamma \nabla f(x[k]), \quad k = 0, 1, 2, \ldots \tag{4.36}$$

to find the minimizer of the quadratic function (4.28).

4.2.4   Proximal Operator for Indicator Functions

The *indicator function* of a non-empty set $C \subset \mathbb{R}^n$ is defined by

$$I_C(x) \triangleq \begin{cases} 0, & \text{if } x \in C, \\ \infty, & \text{if } x \notin C. \end{cases} \tag{4.37}$$

If the set C is non-empty, closed, and convex, then the indicator function $I_C(x)$ is a proper, closed, and convex function (to check this, draw the epigraph). For example, the indicator function $I_C(x)$ of a closed interval C on $\mathbb{R}$ is illustrated in Figure 4.7. You can see that if C is a non-empty closed interval, the epigraph is a non-empty, closed, and convex set.

Let us compute the proximal operator of the indicator function I_C. From the definition (4.19), the proximal operator of I_C is given by

$$\text{prox}_{\gamma I_C}(v) = \arg\min_{x \in \mathbb{R}^n} \left\{ I_C(x) + \frac{1}{2\gamma} \|x - v\|_2^2 \right\}$$

$$= \arg\min_{x \in C} \|x - v\|_2^2 \tag{4.38}$$

$$= \Pi_C(v).$$

That is, the proximal operator of the indicator function I_C is the *projection operator* Π_C onto the set C.

Exercise 4.18. Suppose that $C \subset \mathbb{R}^n$ is a non-empty, closed, and convex set. Prove that $\Pi_C(v)$ is uniquely determined for any $v \in \mathbb{R}^n$.

4.2.5 Proximal Operator for ℓ^1 Norm

Let us compute the proximal operator (4.19) for the ℓ^1 norm:

$$f(x) = \|x\|_1 = \sum_{i=1}^{n} |x_i|, \tag{4.39}$$

where x_i is the i-th element of $x \in \mathbb{R}^n$. From the definition of the proximal operator, we have

$$\text{prox}_{\gamma \|\cdot\|_1}(v) = \arg\min_{x \in \mathbb{R}^n} \left\{ \|x\|_1 + \frac{1}{2\gamma} \|x - v\|_2^2 \right\}$$

$$= \arg\min_{x \in \mathbb{R}^n} \sum_{i=1}^{n} \left\{ |x_i| + \frac{1}{2\gamma} (x_i - v_i)^2 \right\}, \tag{4.40}$$

where v_i is the i-th element of v. This optimization can be reduced to element-wise optimization, that is,

$$\min_{x \in \mathbb{R}^n} \sum_{i=1}^{n} \left\{ |x_i| + \frac{1}{2\gamma} (x_i - v_i)^2 \right\} = \sum_{i=1}^{n} \min_{x_i \in \mathbb{R}} \left\{ |x_i| + \frac{1}{2\gamma} (x_i - v_i)^2 \right\}. \tag{4.41}$$

Therefore, we just solve the following scalar minimization problem:

$$\underset{x \in \mathbb{R}}{\text{minimize}} \ |x| + \frac{1}{2\gamma} (x - v)^2. \tag{4.42}$$

The minimizer $x^* \in \mathbb{R}$ can be easily calculated, which is given by

$$x^* = S_\gamma(v) \triangleq \begin{cases} v - \gamma, & \text{if } v \geq \gamma, \\ 0, & \text{if } -\gamma < v < \gamma, \\ v + \gamma, & \text{if } v \leq -\gamma. \end{cases} \tag{4.43}$$

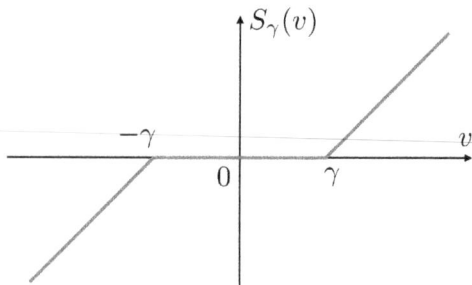

Figure 4.8. Soft-thresholding operator $S_\gamma(v)$.

The function $S_\gamma(v)$ in (4.43) is called the *soft-thresholding operator*. Figure 4.8 shows the graph of the soft-thresholding operator.

Exercise 4.19. Show that the minimizer x^* of the function

$$f(x) \triangleq |x| + \frac{1}{2\gamma}(x - v)^2 \tag{4.44}$$

is given by (4.43). (Hint: divide the domain of $f(x)$ in two intervals: $x \geq 0$ and $x < 0$. Then, consider the three cases for v: $v \geq \gamma$, $-\gamma < v < \gamma$, and $v \leq -\gamma$).

By using the scalar-valued soft-thresholding operator, the proximal operator of the ℓ^1 norm is given by

$$\left[\text{prox}_{\gamma f}(v)\right]_i = S_\gamma(v_i), \tag{4.45}$$

where $[\ \ ]_i$ denotes the i-th element of the vector in the square bracket. For a simple expression, we extend the definition of the scalar-valued soft-thresholding operator (4.43) to vectors. For a vector $v \in \mathbb{R}^n$, we define the vector-valued soft-thresholding operator $S_\gamma(v)$ by

$$[S_\gamma(v)]_i \triangleq S_\gamma(v_i), \tag{4.46}$$

where $[S_\gamma(v)]_i$ is the i-th element of $S_\gamma(v)$. With this notation, the proximal operator of the ℓ^1 norm (4.45) is simply rewritten as

$$\text{prox}_{\gamma \|\cdot\|_1}(v) = S_\gamma(v). \tag{4.47}$$

Exercise 4.20. Let $Q \in \mathbb{R}^{n \times n}$ be an orthogonal matrix. Prove that the minimizer $x^* \in \mathbb{R}^n$ of the following function

$$f(x) \triangleq \frac{1}{2}\|Qx - y\|_2^2 + \lambda\|x\|_1 \tag{4.48}$$

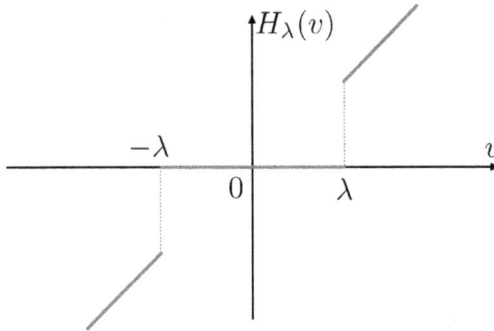

Figure 4.9. Hard-thresholding operator $H_\lambda(v)$.

is given by

$$x^* = S_\lambda(Q^\top y). \tag{4.49}$$

Note that Q is orthogonal if and only if

$$QQ^\top = Q^\top Q = I. \tag{4.50}$$

In summary, the proximal operator of the ℓ^1 norm is the soft-thresholding operator $S_\gamma(v)$. If the absolute value of an element v_i in v is less than γ, then the element is set to be zero by the proximal operator. This is an important property to understand why ℓ^1 optimization gives a sparse solution.

The word 'soft' means that the operator is continuous (see Figure 4.8). We can also define the *hard*-thresholding operator by

$$H_\lambda(v) \triangleq \begin{cases} v, & \text{if } |v| \geq \lambda, \\ 0, & \text{if } |v| < \lambda. \end{cases} \tag{4.51}$$

Figure 4.9 shows the graph of the hard thresholding operator. We can see from this figure, the hard-thresholding operator is discontinuous. An interesting fact is that the hard-thresholding operator is the proximal operator of the ℓ^0 norm with $\lambda = \sqrt{2\gamma}$. Strictly speaking, this is dubious since the proximal operator is defined for proper, closed, and convex functions (see Definition 4.13), but the ℓ^0 norm is not convex. However, the hard-thresholding operator is very useful to derive efficient algorithms for ℓ^0-norm optimization. See Chapter 5 for details.

Exercise 4.21. Compute the proximal operator (4.19) of the ℓ^0 norm, and show that it is the hard-thresholding operator (4.51).

4.3   Proximal Splitting Methods for ℓ^1 Optimization

In this section, we derive an efficient algorithm based on proximal splitting to solve the ℓ^1 optimization:

$$\underset{x \in \mathbb{R}^n}{\text{minimize}} \ \|x\|_1 \quad \text{subject to} \ \Phi x = y, \tag{4.52}$$

where $\Phi \in \mathbb{R}^{m \times n}$ and $y \in \mathbb{R}^m$ are given. We assume that $m < n$ and Φ has full row rank, that is, $\text{rank}(\Phi) = m$. The cost function is the ℓ^1 norm, which is obviously a proper, closed, and convex function. Then, let us consider the constraint. Denote by $\mathcal{C}$ the set of vectors $x \in \mathbb{R}^n$ satisfying the constraint $\Phi x = y$. That is,

$$\mathcal{C} \triangleq \left\{ x \in \mathbb{R}^n : \Phi x = y \right\}. \tag{4.53}$$

It is easy to prove that this set is a non-empty, closed, and convex set in $\mathbb{R}^n$.

Exercise 4.22. Show the cost function $\|x\|_1$ in (4.52) is a proper, closed, and convex function. Also, show the set $\mathcal{C}$ defined in (4.53) is a non-empty, closed, and convex set in $\mathbb{R}^n$.

Then, consider the indicator function $I_{\mathcal{C}}(x)$ for $\mathcal{C}$:

$$I_{\mathcal{C}}(x) = \begin{cases} 0, & \text{if } \Phi x = y, \\ \infty, & \text{if } \Phi x \neq y. \end{cases} \tag{4.54}$$

By using this, the optimization problem in (4.52) is equivalently rewritten as

$$\underset{x \in \mathbb{R}^n}{\text{minimize}} \ \|x\|_1 + I_{\mathcal{C}}(x). \tag{4.55}$$

Note that the functions $\|x\|_1$ and $I_{\mathcal{C}}(x)$ are both proper, closed, and convex functions, and hence the sum of them, $\|x\|_1 + I_{\mathcal{C}}(x)$, is also proper, closed, and convex.

Exercise 4.23. Suppose that two functions, f and g, are proper, closed, and convex. Suppose also that $\text{dom}(f) \cap \text{dom}(g) \neq \emptyset$. Prove that $f + g$ is also a proper, closed, and convex function.

Note that for the optimization problem (4.55), we cannot obtain the proximal operator of the cost function

$$f(x) \triangleq \|x\|_1 + I_{\mathcal{C}}(x), \tag{4.56}$$

in a closed form. In other words, $f(x)$ is not proximable. That is, we cannot directly apply the proximal algorithm (4.25) to this problem. However, the proximal operators of the two functions

$$f_1(x) \triangleq \|x\|_1, \quad f_2(x) \triangleq I_C(x) \tag{4.57}$$

can be obtained as the soft-thresholding operator in (4.47) and the projection operator onto C defined in (4.38), respectively. The idea is to *split* the cost function as $f = f_1 + f_2$, and write an algorithm using the proximal operators of f_1 and f_2 separately. Algorithms designed by this idea are called *proximal splitting algorithms*. For the problem (4.55), we introduce two proximal splitting algorithms.

4.3.1 Douglas-Rachford Splitting Algorithm

Let us consider the following optimization problem in a general form:

$$\underset{x \in \mathbb{R}^n}{\text{minimize}} \ f_1(x) + f_2(x), \tag{4.58}$$

where f_1 and f_2 are proper, closed, and convex functions. The *Douglas-Rachford splitting algorithm* for (4.58) is given as follows:

Douglas-Rachford splitting algorithm for (4.58)

Initialization: give an initial vector $z[0]$ and a parameter $\gamma > 0$
Iteration: for $k = 0, 1, 2, \ldots$ do

$$\begin{aligned} x[k+1] &= \text{prox}_{\gamma f_1}(z[k]) \\ z[k+1] &= z[k] + \text{prox}_{\gamma f_2}(2x[k+1] - z[k]) - x[k+1] \end{aligned} \tag{4.59}$$

From the algorithm, we can derive an algorithm for our unconstrained problem (4.55). In our case, $f_1(x) = \|x\|_1$ and $f_2(x) = I_C(x)$, for which the proximal operators are given by

$$\text{prox}_{\gamma f_1}(v) = S_\gamma(v), \quad \text{prox}_{\gamma f_2}(v) = \Pi_C(v). \tag{4.60}$$

Then the Douglas-Rachford splitting algorithm for the ℓ^1 optimization problem (4.52) is given as follows:

Douglas-Rachford splitting algorithm for ℓ^1 optimization problem (4.52)

Initialization: give an initial vector $z[0]$ and a parameter $\gamma > 0$
Iteration: for $k = 0, 1, 2, \ldots$ do

$$\begin{aligned} x[k+1] &= S_\gamma(z[k]) \\ z[k+1] &= z[k] + \Pi_C(2x[k+1] - z[k]) - x[k+1] \end{aligned} \tag{4.61}$$

In this algorithm, the projection operator Π_C on the linear subspace C defined in (4.53) is given by

$$\Pi_C(v) = v + \Phi^\top(\Phi\Phi^\top)^{-1}(y - \Phi v). \tag{4.62}$$

Note that $\Phi\Phi^\top$ is invertible since Φ has full row rank.

Exercise 4.24. Show that the projection operator Π_C for C defined in (4.53) is given by (4.62).

The ℓ^1 optimization problem (4.52) can be rewritten as a linear programming problem, which can be efficiently solved by the well-known interior-point method [47, Section 5.12.]. However, this method should solve a system of linear equations at each step of the iteration, which takes in general non-negligible computational time. On the other hand, the Douglas-Rachford algorithm in (4.61) only requires

- simple continuous mapping of the soft-thresholding function S_y,
- linear computation of matrix-vector multiplication and vector addition.

Thus, the Douglas-Rachford algorithm is efficient and easy to implement compared to standard interior-point algorithms.

To consider the convergence of Douglas-Rachford splitting algorithm, we define the *relative interior* ri(C) of a subset $C \subset \mathbb{R}^n$ by

$$\text{ri}(C) \triangleq \{x \in \mathbb{R}^n : x \in C \text{ and } \exists \epsilon > 0, \mathcal{N}_\epsilon(x) \cap \text{aff}(C) \subset C\}, \tag{4.63}$$

where $\mathcal{N}_\epsilon(x)$ is the ϵ-neighborhood of x, that is,

$$\mathcal{N}_\epsilon \triangleq \{v \in \mathbb{R}^n : \|v - x\|_2 < \epsilon\}, \tag{4.64}$$

and aff(C) is the affine hull of C, that is, the set of all affine sets containing C. Note that the relative interior is different from the *interior* of C that is defined by

$$\text{int}(C) \triangleq \{x \in \mathbb{R}^n : x \in C \text{ and } \exists \epsilon > 0, \mathcal{N}_\epsilon(x) \subset C\}. \tag{4.65}$$

For example, let us consider the disk

$$C = \{(x_1, x_2, 0) \in \mathbb{R}^3 : x_1^2 + x_2^2 \leq 1\}, \tag{4.66}$$

on the x_1-x_2 plane in $\mathbb{R}^3$. Then, the interior of C is empty by definition, while the relative interior is

$$\text{ri}(C) = \{(x_1, x_2, 0) \in \mathbb{R}^3 : x_1^2 + x_2^2 < 1\}. \tag{4.67}$$

Now, we introduce the convergence theorem [25] for Douglas-Rachford splitting algorithm.

Theorem 4.25. *Suppose that f_1 and f_2 are proper, closed, and convex functions that satisfy*

$$\mathrm{ri}\big(\mathrm{dom}(f_1)\big) \cap \mathrm{ri}\big(\mathrm{dom}(f_2)\big) \neq \emptyset. \tag{4.68}$$

Also, suppose that

$$f_1(\boldsymbol{x}) + f_2(\boldsymbol{x}) \to \infty \ \text{as} \ \|\boldsymbol{x}\|_2 \to \infty. \tag{4.69}$$

Then each sequence $\{\boldsymbol{x}[k]\}_{k=0}^{\infty}$ generated by Douglas-Rachford splitting algorithm converges to a solution of the optimization problem (4.58).

4.3.2 Dykstra-like Splitting Algorithm

Here we consider the following optimization problem:

$$\underset{\boldsymbol{x}\in\mathbb{R}^n}{\text{minimize}} \ f_1(\boldsymbol{x}) + f_2(\boldsymbol{x}) + \frac{1}{2}\|\boldsymbol{x} - \boldsymbol{v}\|_2^2, \tag{4.70}$$

where f_1 and f_2 are proper, closed, and convex functions. We apply Douglas-Rachford splitting algorithm to this optimization by splitting the cost function into f_1 and $f_2 + \frac{1}{2}\|\cdot - \boldsymbol{v}\|_2^2$. Then an algorithm called *Dykstra-like splitting algorithm* is obtained as follows.

Dykstra-like splitting algorithm for (4.70)

Initialization: set $\boldsymbol{x}[0] = \boldsymbol{v}$ and $\boldsymbol{p}[0] = \boldsymbol{q}[0] = \boldsymbol{0}$; give a parameter $\gamma > 0$
Iteration: for $k = 0, 1, 2, \ldots$ do

$$\boldsymbol{z}[k+1] = \mathrm{prox}_{\gamma f_2}(\boldsymbol{x}[k] + \boldsymbol{p}[k])$$

$$\boldsymbol{p}[k+1] = \boldsymbol{x}[k] + \boldsymbol{p}[k] - \boldsymbol{z}[k]$$

$$\boldsymbol{x}[k+1] = \mathrm{prox}_{\gamma f_1}(\boldsymbol{z}[k] + \boldsymbol{q}[k]) \tag{4.71}$$

$$\boldsymbol{q}[k+1] = \boldsymbol{z}[k] + \boldsymbol{q}[k] - \boldsymbol{x}[k+1]$$

An important application of Dykstra-like algorithm is to find the projection of a point onto the intersection of two convex sets $\mathcal{C}_1$ and $\mathcal{C}_2$, namely to find $\Pi_{\mathcal{C}_1 \cap \mathcal{C}_2}(\boldsymbol{v})$. This is done by setting $f_1 = I_{\mathcal{C}_1}$ and $f_2 = I_{\mathcal{C}_2}$, indicator functions defined in (4.37), for the optimization problem (4.70). Since $\mathrm{prox}_{\gamma I_{\mathcal{C}_1}} = \Pi_{\mathcal{C}_1}$ and

$\text{prox}_{\gamma I_{C_2}} = \Pi_{C_2}$, the algorithm is given by

$$z[k+1] = \Pi_{C_1}(x[k] + p[k])$$

$$p[k+1] = x[k] + p[k] - z[k]$$

$$x[k+1] = \Pi_{C_2}(z[k] + q[k])$$ (4.72)

$$q[k+1] = z[k] + q[k] - x[k+1]$$

This is called *Dykstra's projection algorithm*, proposed by Dykstra [14].[1] The name "Dykstra-like splitting" is actually after this algorithm. The convergence theorem is given as follows [25].

Theorem 4.26. *Suppose that f_1 and f_2 are proper, closed, and convex functions that satisfy*

$$\text{dom}(f_1) \cap \text{dom}(f_2) \neq \emptyset.$$ (4.73)

Then each sequence $\{x[k]\}_{k=0}^{\infty}$ generated by Dykstra-like splitting algorithm converges to a solution of the optimization problem (4.70).

Compared to the assumptions in Theorem 4.25 for Douglas-Rachford splitting algorithm, the assumption (4.73) is weaker.

4.4   Proximal Gradient Method for ℓ^1 Regularization

We here consider an efficient algorithm for ℓ^1 regularization (or LASSO):

$$\underset{x \in \mathbb{R}^n}{\text{minimize}} \ \frac{1}{2}\|\Phi x - y\|_2^2 + \lambda\|x\|_1.$$ (4.74)

We assume that $\Phi \in \mathbb{R}^{m \times n}$, $y \in \mathbb{R}^m$, and $\lambda > 0$ are already given.

4.4.1   Algorithm

In (4.74), the first term $\frac{1}{2}\|\Phi x - y\|_2^2$ and the second term $\lambda\|x\|_1$ are both proper, closed, and convex functions of x. Also, the proximal operator of the first term, a quadratic function of x, is obtained in a closed form as described in Section 4.2.3 (see also Exercise 4.27 below). Hence, we can directly apply the Douglas-Rachford splitting algorithm (4.59) to this problem.

---

1.   Note that Dykstra for this algorithm is different from *Dijkstra* who found a famous algorithm for a shortest path in a network.

Exercise 4.27. Show that the proximal operator of $f(x) = \frac{1}{2}\|\Phi x - y\|_2^2$ is given by

$$\text{prox}_{\gamma f}(v) = \left(\Phi^{\top}\Phi + \frac{1}{\gamma}I\right)^{-1}\left(\Phi^{\top}y + \frac{1}{\gamma}v\right). \qquad (4.75)$$

As we have seen before, the proximal operator is an "alternative" to the gradient descent update as shown in (4.35). However, the first term $\frac{1}{2}\|\Phi x - y\|_2^2$ of (4.74) is a quadratic function of x, which is *differentiable*. So, we can directly benefit from the gradient of $\frac{1}{2}\|\Phi x - y\|_2^2$ itself for an algorithm instead of its proximal operator. We here introduce an efficient algorithm using the gradient, which can be much faster than the Douglas-Rachford splitting algorithm, along with an acceleration method.

First, let us consider the following general problem:

$$\underset{x\in\mathbb{R}^n}{\text{minimize}} \ f_1(x) + f_2(x), \qquad (4.76)$$

where f_1 is a differentiable and convex function satisfying $\text{dom}(f_1) = \mathbb{R}^n$, and f_2 is a proper, closed, and convex function. Note that f_2 is not necessarily differentiable, for example an indicator function as in (4.37).

For the optimization problem, we introduce the *proximal gradient algorithm*, which is given as follows:

Proximal gradient algorithm for (4.76)

Initialization: give an initial vector $x[0]$ and a real number $\gamma > 0$
Iteration: for $k = 0, 1, 2, \ldots$ do

$$x[k+1] = \text{prox}_{\gamma f_2}\big(x[k] - \gamma \nabla f_1(x[k])\big). \qquad (4.77)$$

In this algorithm, $\gamma > 0$ is the *step size* of the update. The function $\nabla f_1(x)$ is the gradient of f_1 at $x \in \mathbb{R}^n$.

We offer a geometrical interpretation of the proximal gradient algorithm. Let us define

$$\phi(x) \triangleq \text{prox}_{\gamma f_2}\big(x - \gamma \nabla f_1(x)\big). \qquad (4.78)$$

Then, from the definition of the proximal operator in (4.19), we have

$$\phi(x) = \underset{z\in\mathbb{R}^n}{\arg\min}\left\{ f_2(z) + \frac{1}{2\gamma}\big\|z - (x - \gamma \nabla f_1(x))\big\|_2^2\right\}$$

$$= \underset{z\in\mathbb{R}^n}{\arg\min}\left\{ \tilde{f}_1(z; x) + f_2(z) + \frac{1}{2\gamma}\|z - x\|_2^2\right\}, \qquad (4.79)$$

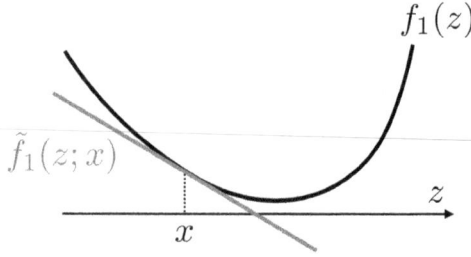

Figure 4.10. Linear approximation $\tilde{f}_1(z; x)$ of convex function $f_1(z)$ at x.

where

$$\tilde{f}_1(z; x) \triangleq f_1(x) + \nabla f_1(x)^\top (z - x). \tag{4.80}$$

Note that $\|\nabla f_1(x)\|_2^2$ and $f_1(x)$ are constant for the minimization with z, and hence $\|\nabla f_1(x)\|_2^2$ is eliminated and $f_1(x)$ is added in (4.79). The function $\tilde{f}_1(z; x)$ is a linear approximation of $f_1(z)$ around the point $x \in \mathbb{R}^n$. Figure 4.10 shows an example of the linear approximation for one dimensional case. From (4.79), the function $\phi(x)$ is the proximal operator of the linearized function $\tilde{f}_1(z; x)$ plus $f_2(z)$, and the iteration (4.77) can be interpreted as the proximal algorithm (4.24) for this approximated function.

4.4.2 Convergence Analysis

Here we analyze the convergence of the proximal gradient algorithm. For this, we define Lipschitz continuity. A function $f : \mathbb{R}^n \to \mathbb{R}^n$ is said to be *Lipschitz continuous* over $\mathbb{R}^n$ if there exists a constant $L > 0$ such that the following inequality

$$\|f(x) - f(y)\|_2 \leq L\|x - y\|_2 \tag{4.81}$$

holds for any vectors $x, y \in \mathbb{R}^n$. When f is Lipschitz continuous, L is called a *Lipschitz constant*, and the smallest L that satisfies (4.81) is called the *best Lipschitz constant*.

Let us consider the optimization problem (4.76). We assume that the gradient ∇f_1 of f_1 is Lipschitz continuous, that is, there exists $L > 0$ such that

$$\|\nabla f_1(x) - \nabla f_1(y)\|_2 \leq L\|x - y\|_2, \quad \forall x, y \in \mathbb{R}^n \tag{4.82}$$

holds. Assume that the optimization problem (4.76) has an optimal solution x^*. Then we have [92, Section 4.2]

$$x^* = \phi(x^*) = \text{prox}_{\gamma f_2}\left(x^* - \gamma \nabla f_1(x^*)\right). \tag{4.83}$$

This implies that an optimal solution x^* of (4.76) is also a *fixed point* of mapping ϕ in (4.78). From this, the meaning of the iteration (4.77) is now clear; this algorithm seeks the fixed point of ϕ.

Exercise 4.28. Consider a continuous function $\phi : \mathbb{R}^n \mapsto \mathbb{R}^n$. Assume that there exists an initial vector $x[0] \in \mathbb{R}^n$ such that the iteration

$$x[k+1] = \phi(x[k]), \quad k = 0, 1, 2, \ldots \tag{4.84}$$

converges to $x^* \in \mathbb{R}^n$. Prove that x^* is a fixed point of mapping ϕ, that is, $x^* = \phi(x^*)$ holds.

In fact, the following theorem holds [5].

Theorem 4.29. *Assume the gradient ∇f_1 is Lipschitz continuous over $\mathbb{R}^n$, and L is a Lipschitz constant satisfying (4.82). Assume also that the step size γ satisfies*

$$\gamma \leq \frac{1}{L}. \tag{4.85}$$

Then the sequence $\{x[k]\}$ generated by the proximal gradient algorithm (4.77) converges to a solution x^* of (4.76), and we have

$$\|x[k+1] - x^*\|_2 \leq \|x[k] - x^*\|_2, \quad k = 0, 1, 2, \ldots \tag{4.86}$$

Moreover, we have

$$f(x[k]) - f(x^*) \leq \frac{L\|x[0] - x^*\|_2^2}{2k}, \quad k = 0, 1, 2, \ldots, \tag{4.87}$$

where $f(x) = f_1(x) + f_2(x)$.

By this theorem, the convergence rate of the proximal gradient algorithm is $O(1/k)$. Note that this rate is much slower than *linear convergence* (or *first-order convergence*), with which the rate is $O(r^k)$ with $|r| < 1$.

Now, let us derive the proximal gradient algorithm of our ℓ^1 regularization (4.74). In our case, the two functions are

$$f_1(x) = \frac{1}{2}\|\Phi x - y\|_2^2, \quad f_2(x) = \lambda\|x\|_1, \tag{4.88}$$

and the gradient of $f_1(x)$ is given by

$$\nabla f_1(x) = \Phi^\top(\Phi x - y). \tag{4.89}$$

Also, the proximal operator of $f_2(x) = \lambda \|x\|_1$ is the soft-thresholding operator (see Section 4.2.5):

$$\text{prox}_{\gamma f_2}(v) = S_{\gamma \lambda}(v). \tag{4.90}$$

From these, the proximal gradient algorithm for (4.74) is given as follows.

Proximal gradient algorithm (ISTA) for (4.74)

Initialization: give an initial vector $x[0]$ and parameter $\gamma > 0$
Iteration: for $k = 0, 1, 2, \ldots$ do

$$x[k+1] = S_{\gamma \lambda}\big(x[k] - \gamma \Phi^\top(\Phi x[k] - y)\big). \tag{4.91}$$

This algorithm is called the *iterative shrinkage thresholding algorithm*, or *ISTA* for short. From (4.89), a Lipschitz constant of ∇f_1 is

$$L = \lambda_{\max}(\Phi^\top \Phi) = \sigma_{\max}(\Phi)^2 = \|\Phi\|^2, \tag{4.92}$$

where $\lambda_{\max}$ and $\sigma_{\max}$ respectively denote the *maximum eigenvalue* and the *maximum singular value*, and $\|\Phi\|$ is a matrix norm defined by $\sigma_{\max}(\Phi)$. Note that if $\Phi \neq 0$, then $\|\Phi\| > 0$. From (4.85) in Theorem 4.29, if we choose γ to satisfy

$$0 < \gamma \leq \frac{1}{\|\Phi\|^2}, \tag{4.93}$$

then a solution of the ℓ^1 regularization (4.74) is obtained after the simple iteration of (4.91).

Theorem 4.29 implies that the error by ISTA decreases at the rate of $O(1/k)$. We can then accelerate the algorithm by using not only $x[k]$ but also the previous $x[k-1]$ in the k-th step. The following algorithm is the accelerated iteration called *FISTA* (Fast ISTA), which converges at the rate of $O(1/k^2)$ [5, 118].

Fast ISTA (FISTA) for (4.74)

Initialization: give initial vectors $x[0]$, $z[0]$, initial number $t[0]$, and parameter $\gamma > 0$
Iteration: for $k = 0, 1, 2, \ldots$ do

$$x[k+1] = S_{\gamma \lambda}\big(z[k] - \gamma \Phi^\top(\Phi z[k] - y)\big),$$

$$t[k+1] = \frac{1 + \sqrt{1 + 4t[k]^2}}{2}, \tag{4.94}$$

$$z[k+1] = x[k+1] + \frac{t[k] - 1}{t[k+1]}(x[k+1] - x[k]).$$

It is surprising that such simple modification leads to improvement of computational efficiency from $O(1/k)$ to $(1/k^2)$. However, it is known that $O(1/k^2)$ is optimal and one cannot accelerate the algorithm any further [92, Section 4.3].

4.5 Generalized LASSO and ADMM

In this section, we consider an extension of ℓ^1 regularization, with a generalized regularization term:

$$\underset{x \in \mathbb{R}^n}{\text{minimize}} \ \frac{1}{2}\|\Phi x - y\|_2^2 + \lambda\|\Psi x\|_1, \qquad (4.95)$$

where Ψ is a matrix. We call this optimization problem the *generalized LASSO*. If Ψ is the identity matrix, this problem is reduced to the ℓ^1 regularization in (4.74). A problem is that the regularization term $\|\Psi x\|_1$ is in general not *proximable*, that is, it is difficult to obtain a closed form of the proximal operator of $\|\Psi x\|_1$. Therefore, we do not directly apply Douglas-Rachford splitting nor the proximal gradient method to this problem. In this section, we introduce an alternative splitting method for this case.

4.5.1 Algorithm

Aside from the generalized LASSO in (4.95), let us consider a general optimization problem:

$$\underset{x \in \mathbb{R}^n, z \in \mathbb{R}^p}{\text{minimize}} \ f_1(x) + f_2(z) \ \text{subject to} \ z = \Psi x, \qquad (4.96)$$

where $f_1 : \mathbb{R}^n \to \mathbb{R} \cup \{\infty\}$ and $f_2 : \mathbb{R}^p \to \mathbb{R} \cup \{\infty\}$ are proper, closed, and convex functions, and $\Psi \in \mathbb{R}^{p \times n}$. The following algorithm, called *Alternating Direction Method of Multipliers*, or *ADMM* for short, is an efficient algorithm to solve (4.96):

ADMM for (4.96)

Initialization: give initial vectors $z[0]$, $v[0] \in \mathbb{R}^p$, and real number $\gamma > 0$
Iteration: for $k = 0, 1, 2, \ldots$ do

$$x[k+1] := \arg\min_{x \in \mathbb{R}^n} \left\{ f_1(x) + \frac{1}{2\gamma}\|\Psi x - z[k] + v[k]\|^2 \right\}, \qquad (4.97)$$

$$z[k+1] := \text{prox}_{\gamma f_2}(\Psi x[k+1] + v[k]), \qquad (4.98)$$

$$v[k+1] := v[k] + \Psi x[k+1] - z[k+1]. \qquad (4.99)$$

To analyze this algorithm, we introduce the *augmented Lagrangian*:

$$L_\rho(x, z, \lambda) = f_1(x) + f_2(z) + \lambda^\top (\Psi x - z) + \frac{\rho}{2}\|\Psi x - z\|_2^2, \quad (4.100)$$

where λ is a Lagrange multiplier and ρ is a positive constant. The term "augmented" means that the function (4.100) is augmented from the usual *Lagrangian*

$$L(x, z, \lambda) = f_1(x) + f_2(z) + \lambda^\top (\Psi x - z). \quad (4.101)$$

by adding the term $\frac{\rho}{2}\|\Psi x - z\|_2^2$. Note that the augmented Lagrangian becomes strongly convex with respect to variables x and z thanks to the additional term $\frac{\rho}{2}\|\Psi x - z\|_2^2$ if $\Psi^\top \Psi$ is positive definite.

Now, let $\gamma \triangleq \rho^{-1}$ and $v[k] \triangleq \gamma \lambda[k]$. Then the ADMM algorithm (4.97)–(4.99) can be rewritten in terms of augmented Lagrangian as

$$x[k + 1] = \arg\min_{x \in \mathbb{R}^n} L_\rho(x, z[k], \lambda[k]) \quad (4.102)$$

$$z[k + 1] = \arg\min_{z \in \mathbb{R}^p} L_\rho(x[k + 1], z, \lambda[k]) \quad (4.103)$$

$$\lambda[k + 1] = \lambda[k] + \rho(\Psi x[k + 1] - z[k + 1]), \quad k = 0, 1, 2, \dots \quad (4.104)$$

Exercise 4.30. Show that the algorithm in (4.102)–(4.104) is equivalent to (4.97)–(4.99) under the transformation $\gamma = \rho^{-1}$, $v[k] = \gamma \lambda[k]$.

The important point of this algorithm is that the optimization for variables x, z, and λ is decoupled. The first step (4.102) is minimization of the augmented Lagrangian for x with fixed z and λ. The second step (4.103) is for z with fixed x and λ. The third step (4.103) is update for λ using obtained x and z.

The following is a convergence theorem for the ADMM algorithm [12, 35]:

Theorem 4.31 (Convergence of ADMM). *Consider the optimization problem in (4.96). Assume that f_1 and f_2 are proper, closed, and convex functions. Assume also that the Lagrangian (4.101) has a saddle point, that is, there exist x^*, z^*, and λ^* such that*

$$L(x^*, z^*, \lambda) \leq L(x^*, z^*, \lambda^*) \leq L(x, z, \lambda^*), \quad \forall x, z, \lambda. \quad (4.105)$$

Then, the ADMM algorithm (4.97)–(4.99) satisfies the following convergence properties:

- *The residual*

$$r[k] \triangleq \Psi x[k] - z[k], \quad k = 0, 1, 2, \dots \quad (4.106)$$

converges to **0** *as* $k \to \infty$. *This implies that the iterates converges to a feasible solution of* (4.96).

- *The objective value* $f_1(x[k]) + f_2(z[k])$ *converges to the optimal value*

$$f^* \triangleq \inf_{\substack{x \in \mathbb{R}^n, z \in \mathbb{R}^p \\ \Psi x = z}} f_1(x) + f_2(z). \tag{4.107}$$

- *If* $\Psi^\top \Psi$ *is positive definite, then the sequence* $\{(x[k], z[k])\}$ *converges to an optimal solution* (x^*, z^*) *of the optimization problem* (4.96).

We can now derive the ADMM algorithm for the generalized LASSO (4.95). First, since $f_1(x) = \frac{1}{2}\|\Phi x - y\|_2^2$, the minimization in (4.97) becomes

$$\arg\min_{x \in \mathbb{R}^n} \left\{ \frac{1}{2}\|\Phi x - y\|_2^2 + \frac{1}{2\gamma}\|\Psi x - z[k] + v[k]\|_2^2 \right\}$$

$$= \left(\Phi^\top \Phi + \frac{1}{\gamma}\Psi^\top \Psi \right)^{-1} \left(\Phi^\top y + \frac{1}{\gamma}\Psi^\top (z[k] - v[k]) \right). \tag{4.108}$$

Exercise 4.32. Prove the equality in (4.108).

Next, since $f_2(x) = \lambda\|x\|_1$, the proximal operator in (4.98) is the soft-thresholding function. In summary, the ADMM algorithm for the generalized LASSO (4.95) is given as follows.

ADMM for generalized LASSO (4.95)

Initialization: give initial vectors $z[0], v[0] \in \mathbb{R}^p$, and real number $\gamma > 0$
Iteration: for $k = 0, 1, 2, \ldots$ do

$$x[k+1] = \left(\Phi^\top \Phi + \frac{1}{\gamma}\Psi^\top \Psi \right)^{-1} \left(\Phi^\top y + \frac{1}{\gamma}\Psi^\top (z[k] - v[k]) \right) \tag{4.109}$$

$$z[k+1] = S_{\gamma\lambda}\left(\Psi x[k+1] + v[k] \right) \tag{4.110}$$

$$v[k+1] = v[k] + \Psi x[k+1] - z[k+1]. \tag{4.111}$$

If we compute the inverse matrix $(\Phi^\top \Phi + \gamma^{-1}\Psi^\top \Psi)^{-1}$ *offline* (i.e. before the iteration), the above ADMM algorithm just includes matrix-vector multiplication, vector addition, and element-wise soft-thresholding. By this property, one can implement this algorithm in a small device and execute very fast. Moreover, if the matrix $\Phi^\top\Phi + \gamma^{-1}\Psi^\top\Psi$ is a *tridiagonal matrix*, the linear equation

$$\left(\Phi^\top\Phi + \frac{1}{\gamma}\Psi^\top\Psi\right)x = v \tag{4.112}$$

with unknown x can be solved in $O(n)$ [41, Section 4.3], and the first step (4.109) can be computed very efficiently.

4.5.2 Total Variation Denoising

Here we consider *total variation denoising* for images, which can achieve noise reduction and edge preserving at the same time. Let us assume we have a noisy image $Y \in \mathbb{R}^{n\times m}$, where each element in Y is the pixel value of the image of size $n \times m$. From 2D image data Y, we pull out each column vector, say $y \in \mathbb{R}^n$, and solve the following optimization problem, one by one:

$$\underset{x\in\mathbb{R}^n}{\text{minimize}} \ \ \|x - y\|_2^2 + \lambda \sum_{i=1}^{n-1} |x_{i+1} - x_i|. \tag{4.113}$$

The first term is the ℓ^2 error between x and y for proximity to the data, while the second term is the ℓ^1 norm of the difference, called the *total variation*, for flatness of the result. The optimization problem (4.113) is a special case of the generalized LASSO (4.95) with $\Phi = I$ (identity matrix) and

$$\Psi = \begin{bmatrix} -1 & 1 & 0 & \cdots & 0 \\ 0 & -1 & 1 & \ddots & \vdots \\ \vdots & \ddots & \ddots & \ddots & 0 \\ 0 & \cdots & 0 & -1 & 1 \end{bmatrix}. \tag{4.114}$$

We can directly exploit ADMM (4.109)–(4.111) for this problem. Moreover, the matrix $\Phi^\top\Phi + \gamma^{-1}\Psi^\top\Psi$ is a tridiagonal matrix, and the algorithm can be executed very fast as mentioned above.

The total variation, which is described by $\|\Psi x\|_1$, can be explained as a convex approximation of the ℓ^0 total variation $\|\Psi x\|_0$. Minimizing the ℓ^0 total variation leads to a sparse difference vector, and hence the optimization result can be maximally flat (i.e., the difference $= 0$ in all but a few pixels). A few nonzero differences come from image edges. That is, we assume that there are just a few edges in an image.

Figure 4.11. Original image (left) and noisy image (right).

Figure 4.12. Total variation denoising with $\lambda = 50$ (left), $\lambda = 100$ (right).

Now, we show the results of total variation denoising. Figure 4.11 shows the original image and noisy image. The noise in the noisy image is so-called *salt-and-pepper noise* with noise density 0.05. Roughly speaking, about 5% of the original pixels are randomly replaced by black or white pixels.

From the noisy image, we remove noise by the total variation denoising. We use the ADMM algorithm with $\gamma = 1$. The maximum number of iterations in ADMM is set to $N = 500$. For the 2-D image, we first run total variation denoising horizontally and then vertically. That is, we run the algorithm twice for one image. Figure 4.12 shows the results of denoising with $\lambda = 50$ and $\lambda = 100$. If you take larger λ, the variation between adjacent pixels will be smaller. Comparing images with $\lambda = 50$ and $\lambda = 100$ in Figure 4.12, the image with $\lambda = 100$ gives an impression of more smoothness than that with $\lambda = 50$. This effect is much more

Restored image

Figure 4.13. Result of total variation denoising with $\lambda = 200$.

perceptible when $\lambda = 200$. Figure 4.13 shows the result. The total variation term is too strong in this case, and the restored image is now unacceptable.

In summary, to obtain a good result, you should carefully choose the parameter λ, which affects the quality of denoising. However, there is no general rule for optimal λ, and you should choose λ by trial and error.

MATLAB programs to do this example are available at the end of this chapter. You can experiment the total variation denoising by yourself.

4.6   Further Reading

To study convex optimization deeply, you can choose a renowned book by Boyd and Vandenberghe [13]. The PDF version of the book, lecture slides, and MATALB/Python programs for exercises can be available in

http://web.stanford.edu/~boyd/cvxbook/

You can also choose a recent book by Bertsekas [7]. This book devotes much spaces to recent algorithms such as the proximal gradient algorithms and ADMM. If you need deep and mathematical knowledge of convex operation at a research level, you can consult the book [4] by Bauschke and Combettes. For proximal splitting algorithm, you can refer to [25, 92]. The book chapter [5] by Beck and Teboulle is a nice introduction of ISTA and FISTA. For ADMM, you can read the book [12] by Boyd et al.

MATLAB Programs

To run the main program, you need Image Processing Toolbox in MATLAB.

MATLAB program for Section 4.5.2 (Total variation denoising)

```
%% Read image
Img = imread('cat.jpg'); % read image
X_orig = rgb2gray(Img); % Color to gray
[n,m] = size(X_orig); % Image size

%% Noise (salt & pepper)
Y = imnoise(X_orig,'salt & pepper',0.05);

%% Display images
figure;
imshow(X_orig);
title('Original image');
figure;
imshow(Y);
title('Noisy image');

%% Denoising
% optimization parameter
lambda = 50;
% matrix Phi and Psi
Phi = eye(n);
Psi = -diag(ones(n,1))+diag(ones(n-1,1),1);

% ADMM iteration
gamma = 1; % step size parameter
N = 500; % number of iterations
X_res = zeros(n,m); % restored image
Z = zeros(n,m); V = zeros(n,m); % initial values
M = sparse(Phi'*Phi + (1/gamma)*Psi'*Psi); % sparse matrix

% vertical processing
Yv = double(Y);
W = Phi'*Yv;
for k=1:N
    X = M\(W+gamma\Psi'*(Z-V));
    P = Psi*X+V;
    Z = soft_thresholding(gamma*lambda,P);
    V = P - Z;
end

% horizontal processing
W = rot90(X);
for k = 1:N
    X = M\(W+gamma\Psi'*(Z-V));
    P = Psi*X+V;
    Z = soft_thresholding(gamma*lambda,P);
    V = P - Z;
end
X = rot90(X,3);

%% Result
figure;
imshow(uint8(round(X)));
title('Restored image');
```

MATLAB function for Soft-thresholding operator $S_\lambda(v)$

```
function sv = soft_thresholding(lambda,v)

[m,n] = size(v);
mn = m*n;
sv = zeros(m,n);
for i = 1:mn
  if abs(v(i))<=lambda
    sv(i) = 0;
  else
    sv(i) = v(i) - sign(v(i))*lambda;
  end
end
```

DOI: 10.1561/9781680837254.ch5

Chapter 5

Greedy Algorithms

In the previous chapter, we have formulated the problem of sparse representation as optimization problems with ℓ^1 norm, for which there are efficient and fast algorithms. The idea was to approximate the non-convex and discontinuous ℓ^0 norm by the convex ℓ^1 norm. In this chapter, we consider alternative algorithms that directly solve the ℓ^0-norm optimization by using the greedy method.

Key ideas of Chapter 5

- Greedy algorithms are available to directly solve ℓ^0 optimization.
- The greedy algorithms introduced in this chapter show the linear convergence, which are much faster than the proximal splitting algorithms.
- A local optimal solution is obtained by a greedy algorithm, which is not necessarily a global optimizer.

5.1 ℓ^0 Optimization

Let us consider the following ℓ^0 optimization problem:

$$\underset{x \in \mathbb{R}^n}{\text{minimize}} \quad \|x\|_0 \quad \text{subject to} \quad \Phi x = y, \tag{5.1}$$

where we assume $\Phi \in \mathbb{R}^{m \times n}$ and $y \in \mathbb{R}^m$ are given. To consider this optimization problem, let us first define the *mutual coherence* of a matrix.

Definition 5.1. *For a matrix* $\Phi = [\phi_1, \phi_2, \ldots, \phi_n] \in \mathbb{R}^{m \times n}$ *with* $\phi_i \in \mathbb{R}^m$, $i = 1, 2, \ldots, n$, *we define the* mutual coherence $\mu(\Phi)$ *by*

$$\mu(\Phi) \triangleq \max_{\substack{i,j=1,\ldots,n \\ i \neq j}} \frac{|\langle \phi_i, \phi_j \rangle|}{\|\phi_i\|_2 \|\phi_j\|_2}. \tag{5.2}$$

The mutual coherence is the maximum value of the absolute value of the inner product of $\phi_i/\|\phi_i\|_2$ and $\phi_j/\|\phi_j\|_2$. That is,

$$\mu(\Phi) = \max_{\substack{i,j=1,\ldots,n \\ i \neq j}} \left| \left\langle \frac{\phi_i}{\|\phi_i\|_2}, \frac{\phi_j}{\|\phi_j\|_2} \right\rangle \right| \tag{5.3}$$

The value $\frac{\langle \phi_i, \phi_j \rangle}{\|\phi_i\|_2 \|\phi_j\|_2}$ is the *cosine* of the angle θ_{ij} between two lines along with ϕ_i and ϕ_j. If the angle is small (i.e. coherent), then this value is large (close to 1), and if the angle is large (close to 90°, incoherent), then the value is almost 0. Figure 5.1 illustrates these properties. Hence, the mutual coherence is described as

$$\mu(\Phi) = \max_{\substack{i,j=1,\ldots,n \\ i \neq j}} |\cos \theta_{ij}|. \tag{5.4}$$

Roughly speaking, if the vectors $\phi_1, \ldots, \phi_n$ are uniformly spread in $\mathbb{R}^m$, then the mutual coherence $\mu(\Phi)$ is small. On the other hand, if some vectors in the dictionary are coherent like a tassel, then $\mu(\Phi)$ is large. Figure 5.2 shows dictionaries with large $\mu(\Phi)$ and small one.

From Cauchy-Schwartz inequality

$$|\langle x, y \rangle| \leq \|x\|_2 \|y\|_2, \quad \forall x, y \in \mathbb{R}^m, \tag{5.5}$$

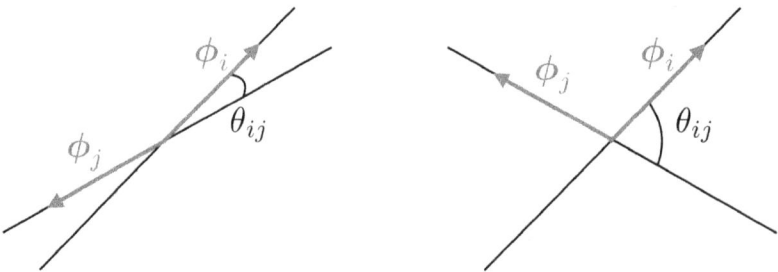

Figure 5.1. Angle θ_{ij} between two lines along with ϕ_i and ϕ_j: coherent vectors (left) and incoherent vectors (right).

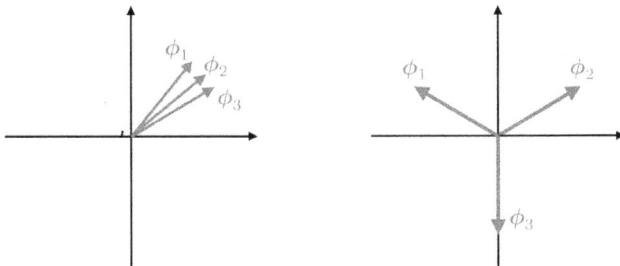

Figure 5.2. Dictionary $\{\phi_1, \phi_2, \phi_3\}$ with large $\mu(\Phi)$ (left) and small $\mu(\Phi)$ (right).

the maximum value of the mutual coherence is 1. Since the equality in (5.5) holds if and only if the two vectors x and y are parallel, we have $\mu(\Phi) = 1$ if and only if there exist parallel vectors in $\{\phi_1, \phi_2, \dots, \phi_n\}$. On the other hand, the mutual coherence is always non-negative, and $\mu(\Phi) = 0$ if Φ is a square orthogonal matrix.

By using the mutual coherence, we can characterize the solution of the ℓ^0 optimization (5.1) [37, Theorem 2.5]:

Theorem 5.2. *If there exists a vector $x \in \mathbb{R}^n$ that satisfies linear equation $\Phi x = y$, and*

$$\|x\|_0 < \frac{1}{2}\left(1 + \frac{1}{\mu(\Phi)}\right), \tag{5.6}$$

then x is the sparsest solution of the linear equation.

By this theorem, let us consider properties of the solution(s) of the ℓ^0 optimization problem in (5.1).

First, let us assume $\mu(\Phi) < 1$. That is, there are no parallel vectors in $\{\phi_1, \phi_2, \dots, \phi_n\}$. Then, we have

$$\frac{1}{2}\left(1 + \frac{1}{\mu(\Phi)}\right) > 1, \tag{5.7}$$

and hence if there exists a 1-sparse solution x (i.e. $\|x\|_0 = 1$) of equation $\Phi x = y$, then this is the sparsest solution from Theorem 5.2. Now, we have

$$y = \Phi x = x_1\phi_1 + x_2\phi_2 + \cdots + x_n\phi_n, \tag{5.8}$$

and hence the 1-sparse solution is parallel to one of $\phi_1, \phi_2, \dots, \phi_n$. From this, we find ϕ_i that is parallel to y. This is formulated as a problem to find an index $i \in \{1, 2, \dots, n\}$ that minimizes the error $e(i)$ defined by

$$e(i) \triangleq \min_{x \in \mathbb{R}} \|x\phi_i - y\|_2^2. \tag{5.9}$$

If there exists x with $\|x\|_0 = 1$, then there exists an index $i \in \{1, 2, \ldots, n\}$ that achieves $e(i) = 0$. Now, the minimum value of (5.9) can be easily obtained as follows:

$$
\begin{aligned}
e(i) &= \min_{x \in \mathbb{R}} \|x\phi_i - y\|_2^2 \\
&= \min_{x \in \mathbb{R}} \left\{ \langle \phi_i, \phi_i \rangle x^2 - 2\langle \phi_i, y \rangle x + \langle y, y \rangle \right\} \\
&= \min_{x \in \mathbb{R}} \left\{ \|\phi_i\|_2^2 \left(x - \frac{\langle \phi_i, y \rangle}{\|\phi_i\|_2^2} \right)^2 + \|y\|_2^2 - \frac{\langle \phi_i, y \rangle^2}{\|\phi_i\|_2^2} \right\} \\
&= \|y\|_2^2 - \frac{\langle \phi_i, y \rangle^2}{\|\phi_i\|_2^2}.
\end{aligned}
\tag{5.10}
$$

From this formula, we can find one index i^* that satisfies $e(i^*) = 0$ (if it exists) by computing $e(i)$ for $i = 1, 2, \ldots, n$. Then we have

$$
y = x^* \phi_{i^*}, \quad x^* \triangleq \frac{\langle \phi_{i^*}, y \rangle}{\|\phi_{i^*}\|_2^2},
\tag{5.11}
$$

and the corresponding 1-sparse vector x^* is given by

$$
x^* = [0, \ldots, 0, \overset{\overset{i^*}{\vee}}{x^*}, 0, \ldots, 0]^\top.
\tag{5.12}
$$

This computation requires $O(n)$ computational time at the worst case.

Let us generalize this observation. Assume that there exists a natural number k that satisfies

$$
\mu(\Phi) < \frac{1}{2k - 1}.
\tag{5.13}
$$

Then we have

$$
\frac{1}{2}\left(1 + \frac{1}{\mu(\Phi)} \right) > \frac{1}{2}(1 + 2k - 1) = k.
\tag{5.14}
$$

Assume also that there exists a k-sparse solution (i.e. $\|x\|_0 \le k$) of the linear equation $\Phi x = y$. From Theorem 5.2, this is the sparsest solution. Then, the vector y is a linear combination of k vectors in the dictionary $\{\phi_1, \phi_2, \ldots, \phi_n\}$. As we have seen in Section 2.4 in Chapter 2 (p. 24), to find the k-sparse solution by the exhaustive search, we need $\binom{n}{k}$ or $O(n^k)$ computations, which cannot acceptable in large scale problems.

For such problems, a method called the *greedy method* is available. This method is an iterative method for a global solution, in which the locally optimal choice is made at each stage. Although this method does not always give a global solution, this method is a powerful tool for combinatorial problems. In the next section, we introduce greedy algorithms for the ℓ^0 optimization problem in (5.1).

5.2 Orthogonal Matching Pursuit

5.2.1 Matching Pursuit (MP)

First, we introduce the simplest greedy algorithm called *matching pursuit* (MP for short) to solve the ℓ^0 optimization problem in (5.1). This algorithm iteratively seeks a 1-sparse vector that is a solution of a local ℓ^0 optimization problem. As mentioned above, a 1-sparse optimal vector is easily obtained with $O(n)$ computations. Matching pursuit aims at finding a global solution by iterating such an easy local optimization problem.

The algorithm of matching pursuit iteratively approximates the solution of linear equation $\Phi x = y$ by decreasing the *residual* $r[k] = y - \Phi x[k]$ at each step. The procedure is shown as follows:

1. Find a 1-sparse vector $x[1]$ that minimizes $\|\Phi x - y\|_2$.
2. For $k = 1, 2, 3, \ldots$ do
   - Compute the residual $r[k] = y - \Phi x[k]$
   - Find a 1-sparse vector x^* that minimizes $\|\Phi x - r[k]\|_2$ and set

$$x[k+1] = x[k] + x^*.$$

At the first step, we seek a 1-sparse vector $x[1]$ that minimizes $\|\Phi x - y\|_2$. Let $x[1]$ be the non-zero element of $x[1]$ and $i[k]$ the corresponding index, that is,

$$x[1] = \left[0, \ldots, 0, \overset{i[1]}{\overset{\vee}{x[1]}}, 0, \ldots, 0\right]^\top = x[1] e_{i[1]}, \tag{5.15}$$

where $e_i, i \in \{1, \ldots, n\}$ is the standard basis in $\mathbb{R}^n$ defined by

$$e_i \triangleq \left[0, \ldots, 0, \overset{i}{\overset{\vee}{1}}, 0, \ldots, 0\right]^\top \in \mathbb{R}^n. \tag{5.16}$$

Then, from (5.10), $i[1]$ and $x[1]$ are easily obtained as

$$i[1] = \underset{i \in \{1,\ldots,n\}}{\arg\min} \, e(i)$$

$$= \underset{i \in \{1,\ldots,n\}}{\arg\min} \left\{ \|y\|_2^2 - \frac{\langle \phi_i, y \rangle^2}{\|\phi_i\|_2^2} \right\}$$

$$= \arg\max_{i \in \{1,\ldots,n\}} \frac{\langle \phi_i, y \rangle^2}{\|\phi_i\|_2^2},$$

$$x[1] = \frac{\langle \phi_{i[1]}, y \rangle}{\|\phi_{i[1]}\|_2^2}.$$

(5.17)

The residual is given by

$$r[1] = y - \Phi x[1] = y - x[1]\phi_{i[1]},$$

(5.18)

and we have

$$y = x[1]\phi_{i[1]} + r[1].$$

(5.19)

We can easily check (see Exercise 5.3 below) that the residual vector $r[1]$ is orthogonal to $\phi_{i[1]}$, and hence we have

$$\|y\|_2^2 = \|x[1]\phi_{i[1]}\|_2^2 + \|r[1]\|_2^2.$$

(5.20)

If the residual $\|r[1]\|_2$ is sufficiently small, then

$$\tilde{y}[1] \triangleq x[1]\phi_{i[1]} = \Phi x[1]$$

(5.21)

is a good approximation of y. Figure 5.3 illustrates this observation.

Exercise 5.3. Prove that the residual vector $r[1]$ is orthogonal to $\phi_{i[1]}$. Also prove that the equation (5.20) holds.

At the second step, we seek a 1-sparse vector that is the best approximation of the residual vector $r[1]$ in (5.18). The 1-sparse vector is easily obtained by (5.10) with $r[1]$ instead of y. Let $x[2]$ be the 1-sparse vector, $x[2]$ its non-zero element, and $i[2]$ the corresponding index. Then we have

$$i[2] = \arg\max_{i \in \{1,\ldots,n\}} \frac{\langle \phi_i, r[1] \rangle^2}{\|\phi_i\|_2^2}, \quad x[2] = \frac{\langle \phi_{i[2]}, r[1] \rangle}{\|\phi_{i[2]}\|_2^2}.$$

(5.22)

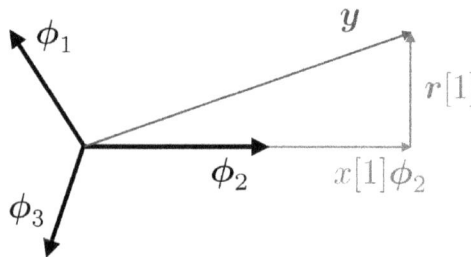

Figure 5.3. Vector $\tilde{y}[1] = x[1]\phi_2$ with $i[1] = 2$ that is the best 1-sparse approximation of y. The residual vector $r[1]$ is orthogonal to ϕ_2.

The residual vector $r[2]$ is given by

$$r[2] = r[1] - \Phi x[2] = r[1] - x[2]\phi_{i[2]}, \tag{5.23}$$

and from (5.19), we have

$$y = x[1]\phi_{i[1]} + x[2]\phi_{i[2]} + r[2]. \tag{5.24}$$

It is easily shown that $\phi_{i[2]}$ and $r[2]$ are orthogonal to each other, and

$$\|r[1]\|_2^2 = \|x[2]\phi_{i[2]}\|_2^2 + \|r[2]\|_2^2. \tag{5.25}$$

From this with (5.20), we have

$$\|y\|_2^2 = \|x[1]\phi_{i[1]}\|_2^2 + \|x[2]\phi_{i[2]}\|_2^2 + \|r[2]\|_2^2. \tag{5.26}$$

Now we obtain a 2-sparse vector

$$x[2] \triangleq [0, \ldots, 0, \overset{i[1]}{\underset{\vee}{x[1]}}, 0, \ldots, 0, \overset{i[2]}{\underset{\vee}{x[2]}}, 0, \ldots, 0]^\top = x[1]e_{i[1]} + x[2]e_{i[2]}, \tag{5.27}$$

which gives a 2-sparse approximation of y as

$$\tilde{y}[2] \triangleq x[1]\phi_{i[1]} + x[2]\phi_{i[2]} = \Phi x[2]. \tag{5.28}$$

Figure 5.4 illustrates this.

If we continue the same procedure, we have at the k-th step

$$y = x[1]\phi_{i[1]} + x[2]\phi_{i[2]} + \cdots + x[k]\phi_{i[k]} + r[k]. \tag{5.29}$$

Define the k-sparse vector by

$$x[k] \triangleq x[1]e_{i[1]} + x[2]e_{i[2]} + \cdots + x[k]e_{i[k]}. \tag{5.30}$$

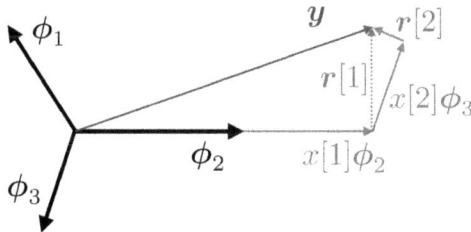

Figure 5.4. Vector $x[2]\phi_3$ with $i[2] = 3$ that is the best 1-sparse approximation of the residual vector $r[1]$. The residual vector $r[2]$ is orthogonal to ϕ_3, and $y = x[1]\phi_2 + x[2]\phi_3 + r[2]$ holds.

Then the vector y is approximated by using this k-sparse vector as

$$\tilde{y}[k] \triangleq x[1]\phi_{i[1]} + x[2]\phi_{i[2]} + \cdots + x[k]\phi_{i[k]} = \Phi x[k]. \qquad (5.31)$$

If the residual $r[k]$ approaches to 0 as $k \to \infty$, then we obtain an approximated solution of the ℓ^0 optimization problem in (5.1) by stopping the procedure at small k. The exhaustive search requires $O(n^k)$ computations to obtain k-sparse vector, while matching pursuit needs just $O(nk)$ computations.

We summarize the algorithm of matching pursuit as follows:

MP for ℓ^0 optimization (5.1)

Initialization: Set $x[0] = 0$, $r[0] = y$, $k = 1$
Iteration: while $\|r[k]\|_2 \geq$ eps, do

$$i[k] := \arg\max_{i \in \{1,\ldots,n\}} \frac{\langle \phi_i, r[k-1] \rangle^2}{\|\phi_i\|_2^2},$$

$$x[k] := \frac{\langle \phi_{i[k]}, r[k-1] \rangle}{\|\phi_{i[k]}\|_2^2},$$

$$\qquad\qquad\qquad (5.32)$$

$$x[k] := x[k-1] + x[k]e_{i[k]},$$

$$r[k] := r[k-1] - x[k]\phi_{i[k]},$$

$$k := k+1$$

In this algorithm, eps is the *termination tolerance* that should be fixed beforehand.

Exercise 5.4. Prove that the following equality holds at the k-th step in the MP algorithm:

$$\|y\|_2^2 = \sum_{j=1}^{k} \|x[j]\phi_{i[j]}\|_2^2 + \|r[k]\|_2^2. \qquad (5.33)$$

Moreover, show that if $\phi_1, \phi_2, \ldots, \phi_n$ are normalized, that is

$$\|\phi_i\|_2 = 1, \quad \forall i \in \{1, 2, \ldots, n\}, \qquad (5.34)$$

then the following equality holds:

$$\|y\|_2^2 = \sum_{j=1}^{k} |x[j]|^2 + \|r[k]\|_2^2. \qquad (5.35)$$

The following theorem gives the convergence property of the MP algorithm [71].

Theorem 5.5. *Assume that dictionary* $\{\phi_1, \phi_2, \ldots, \phi_n\}$ *has* m *linearly independent vectors (i.e.* rank$(\Phi) = m$*). Then there exists a constant* $c \in (0, 1)$ *such that*

$$\|r[k]\|_2^2 \leq c^k \|y\|_2^2, \quad k = 0, 1, 2, \ldots. \tag{5.36}$$

From this theorem, it follows that the residual $r[k]$ monotonically decreases and

$$\lim_{k \to \infty} r[k] = 0 \tag{5.37}$$

holds.

The convergence rate in (5.36) is *first order* or *linear*, and the residual decreases exponentially, that is, $O(c^k)$. This rate is much faster than FISTA in (4.94) (p. 80) for the ℓ^1 regularization, which has $O(1/k^2)$ convergence.

5.2.2   Orthogonal Matching Pursuit (OMP)

We have seen that the residual $r[k]$ by the matching pursuit (MP) algorithm (5.32) decreases very fast. However, in general, it does not always achieve $r[k] = 0$ in a finite number of iterations, and the output vector $x[k]$ for large k, or $\lim_{k\to\infty} x[k]$ may not be sparse. This is because MP may choose an index $i[k]$ that was already chosen in previous steps. *Orthogonal Matching Pursuit* (OMP) is an algorithm to improve MP to achieve a finite number of iterations to obtain a sparse solution. This is done by removing an index from candidates if it was once chosen. Let us see the procedure of OMP precisely.

At the k-th step in MP, we choose the index by

$$i[k] = \underset{i \in \{1, \ldots, n\}}{\arg \max} \frac{\langle \phi_i, r[k-1] \rangle^2}{\|\phi_i\|_2^2}, \quad r[0] = y, \quad k = 1, 2, \ldots \tag{5.38}$$

To memorize indices that were chosen in the previous steps, we define the set $\mathcal{S}_k$ of the chosen indices by k-th step as

$$\mathcal{S}_k = \mathcal{S}_{k-1} \cup \{i[k]\}, \quad \mathcal{S}_0 = \emptyset, \quad k = 1, 2, \ldots \tag{5.39}$$

Also, let us define a linear subspace $\mathcal{C}_k$ of $\mathbb{R}^m$ spanned by vectors $\phi_i, i \in \mathcal{S}_k$, that is,

$$\mathcal{C}_k \triangleq \text{span}\{\phi_i : i \in \mathcal{S}_k\} = \left\{ \sum_{i \in \mathcal{S}_k} x_i \phi_i : x_i \in \mathbb{R} \right\}. \tag{5.40}$$

OMP approximates the vector y at each step by a vector in $\mathcal{C}_k$, while MP approximates it by just one vector $\phi_{i[k]}$. More precisely, OMP chooses a vector $\tilde{y}[k]$ in

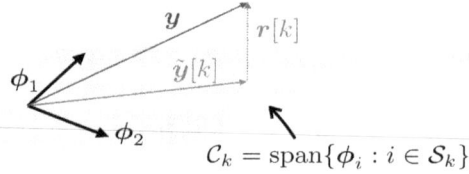

Figure 5.5. The k-th step of OMP: find the best approximation $\tilde{y}[k]$ of y in the linear subspace $\mathcal{C}_k = \text{span}\{\phi_i : i \in \mathcal{S}_k\}$. The residual vector $r[k] = y - \tilde{y}[k]$ is orthogonal to $\mathcal{C}_k$.

$\mathcal{C}_k$ that has the minimum ℓ^2 distance from y. This is obtained by the orthogonal projection of y onto $\mathcal{C}_k$:

$$\tilde{y}[k] = \arg\min_{v \in \mathcal{C}_k} \|v - y\|_2^2 = \Pi_{\mathcal{C}_k}(y), \tag{5.41}$$

where $\Pi_{\mathcal{C}_k}$ is the projection operator onto $\mathcal{C}_k$. Figure 5.5 illustrates this projection at the k-th step.

Using the restriction notation,[1] we can characterize the condition $v \in \mathcal{C}_k$ as

$$v = \sum_{i \in \mathcal{S}_k} x_i \phi_i = \Phi_{\mathcal{S}_k} \tilde{x}, \tag{5.42}$$

for some $\tilde{x} \in \mathbb{R}^k$. Note that $\#(\mathcal{S}_k) = k$ holds as explained later. Then, the projection in (5.41) is obtained by finding the coefficients of $\tilde{y}[k]$ with respect to the basis functions $\phi_i, i \in \mathcal{S}_k$ in $\mathcal{C}_k$. That is, we find

$$\tilde{x}[k] = \arg\min_{\tilde{x} \in \mathbb{R}^k} \frac{1}{2} \|\Phi_{\mathcal{S}_k} \tilde{x} - y\|_2^2. \tag{5.43}$$

This is the least squares solution[2]

$$\tilde{x}[k] = \left(\Phi_{\mathcal{S}_k}^\top \Phi_{\mathcal{S}_k}\right)^{-1} \Phi_{\mathcal{S}_k}^\top y. \tag{5.44}$$

Note that the matrix $\Phi_{\mathcal{S}_k}^\top \Phi_{\mathcal{S}_k}$ is always invertible (this will be explained later). Then $\tilde{y}[k]$ is given by

$$\tilde{y}[k] = \Phi_{\mathcal{S}_k} \tilde{x}[k] = \Phi_{\mathcal{S}_k} \left(\Phi_{\mathcal{S}_k}^\top \Phi_{\mathcal{S}_k}\right)^{-1} \Phi_{\mathcal{S}_k}^\top y. \tag{5.45}$$

Define the coefficient vector $x[k] \in \mathbb{R}^n$ with respect to $\phi_i, i \in \{1, 2, \ldots, n\}$ by

$$\left(x[k]\right)_{\mathcal{S}_k} = \tilde{x}[k], \quad \left(x[k]\right)_{\mathcal{S}_k^c} = 0, \tag{5.46}$$

---

1. For the restriction notation, see (2.45), (2.46), and (2.47) in Chapter 2 (p. 25).

2. For the least squares solution, see Section 3.1.2 in Chapter 3 and equation (3.23).

where $\mathcal{S}_k^c$ is the complement of $\mathcal{S}_k$. Then we have

$$\tilde{y}[k] = \Phi x[k]. \tag{5.47}$$

The residual vector $r[k] = y - \tilde{y}[k]$ is given by

$$r[k] = y - \tilde{y}[k] = \{I - \Phi_{\mathcal{S}_k}(\Phi_{\mathcal{S}_k}^\top \Phi_{\mathcal{S}_k})^{-1}\Phi_{\mathcal{S}_k}^\top\}y. \tag{5.48}$$

It is easily shown that the residual vector $r[k]$ is orthogonal to the linear subspace $\mathcal{C}_k$ (see Figure 5.5), that is,

$$\langle v, r[k] \rangle = 0, \quad \forall v \in \mathcal{C}_k. \tag{5.49}$$

This means that any vector ϕ_i in $\mathcal{C}_k$ will never be chosen by the maximization at the $(k+1)$-th step:

$$i[k+1] = \underset{i \in \{1,2,\ldots,n\}}{\arg\max} \frac{\langle \phi_i, r[k] \rangle^2}{\|\phi_i\|_2^2} = \underset{\substack{i \in \{1,2,\ldots,n\} \\ \phi_i \notin \mathcal{C}_k}}{\arg\max} \frac{\langle \phi_i, r[k] \rangle^2}{\|\phi_i\|_2^2} \tag{5.50}$$

since $\langle \phi_i, r[k] \rangle = 0$ holds for any $\phi_i \in \mathcal{C}_k$, from (5.49). Also, we see that ϕ_i, $i \in \mathcal{S}_k$ are always linearly independent since $\phi_{i[k+1]} \notin \mathcal{C}_k$ holds for any k, and hence $\Phi_{\mathcal{S}_k}^\top \Phi_{\mathcal{S}_k}$ is invertible. The name *orthogonal* matching pursuit comes from this property of orthogonality.

We summarize the algorithm of OMP as follows.

OMP for ℓ^0 optimization (5.1)

Initialization: Set $x[0] = 0$, $r[0] = y$, $\mathcal{S}_0 = \emptyset$, $k = 1$
Iteration: while $r[k] \neq 0$ do

$$i[k] := \underset{i \in \{1,\ldots,n\}}{\arg\max} \frac{\langle \phi_i, r[k-1] \rangle^2}{\|\phi_i\|_2^2},$$

$$\mathcal{S}_k := \mathcal{S}_{k-1} \cup \{i[k]\},$$

$$\tilde{x}[k] := (\Phi_{\mathcal{S}_k}^\top \Phi_{\mathcal{S}_k})^{-1}\Phi_{\mathcal{S}_k}^\top y,$$

$$(x[k])_{\mathcal{S}_k} := \tilde{x}[k], \tag{5.51}$$

$$(x[k])_{\mathcal{S}_k^c} := 0,$$

$$r[k] := y - \Phi_{\mathcal{S}_k}\tilde{x}[k],$$

$$k := k+1$$

The following theorem shows that if there exists a sufficiently sparse solution of the equation $\Phi x = y$, then OMP gives the solution of the ℓ^0 optimization (5.1) in a finite number of iterations [37, Theorem 4.3]:

Theorem 5.6. *Assume that $\Phi \in \mathbb{R}^{m \times n}$ is surjective, that is, $\mathrm{rank}(\Phi) = m$. Assume also that there exists a vector $x \in \mathbb{R}^n$ such that $\Phi x = y$ and*

$$\|x\|_0 < \frac{1}{2}\left(1 + \frac{1}{\mu(\Phi)}\right), \tag{5.52}$$

where $\mu(\Phi)$ is the mutual coherence of matrix Φ. Then, this vector x is the unique solution of the ℓ^0 optimization (5.1), and OMP gives it in $k = \|x\|_0$ steps.

We should note that at each step of OMP we need to compute the matrix inversion of $(\Phi_{\mathcal{S}_k}^\top \Phi_{\mathcal{S}_k})^{-1}\Phi_{\mathcal{S}_k}^\top y$. If the number $k = \|x\|_0$ in Theorem 5.6 is very large, then this inversion may impose a heavy computational burden.

5.3 Thresholding Algorithm

In this section, we consider the following optimization problems:

$$\underset{x \in \mathbb{R}^n}{\text{minimize}} \ \frac{1}{2}\|\Phi x - y\|_2^2 + \lambda\|x\|_0 \tag{5.53}$$

$$\underset{x \in \mathbb{R}^n}{\text{minimize}} \ \frac{1}{2}\|\Phi x - y\|_2^2 \ \text{subject to} \ \|x\|_0 \leq s \tag{5.54}$$

The first problem (5.53) is called the ℓ^0 *regularization*, and the second problem (5.54) is called the *s-sparse approximation*. Note that these optimization problems are non-convex and combinatorial. For these problems, we introduce efficient greedy algorithms by borrowing the idea of the proximal gradient algorithm studied in Chapter 4.

5.3.1 Iterative Hard-thresholding Algorithm (IHT)

Let us consider the following optimization problem:

$$\underset{x \in \mathbb{R}^n}{\text{minimize}} \ f_1(x) + f_2(x), \tag{5.55}$$

where f_1 is a differentiable and convex function satisfying $\mathrm{dom}(f_1) = \mathbb{R}^n$, and f_2 is a proper, closed, and convex function. The proximal gradient algorithm for this

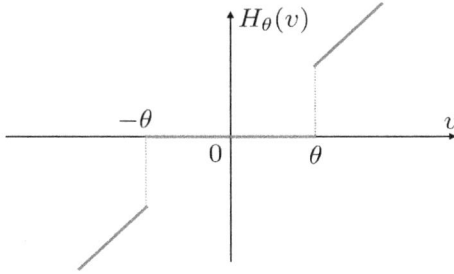

Figure 5.6. Hard-thresholding operator $H_\theta(v)$.

is given by[3]

$$x[k+1] = \text{prox}_{\gamma f_2}\big(x[k] - \gamma \nabla f_1(x[k])\big). \tag{5.56}$$

For the ℓ^0 regularization of (5.53), we have

$$f_1(x) \triangleq \frac{1}{2}\|\Phi x - y\|_2^2, \quad f_2(x) \triangleq \lambda\|x\|_0. \tag{5.57}$$

Although the function f_2 is not convex in this case, we thoughtlessly apply it to the proximal gradient algorithm (5.56). Now, the proximal operator of $f_2(x) = \lambda\|x\|_0$ has a closed form, namely the *hard-thresholding operator* (see Figure 5.6) defined by

$$[H_\theta(v)]_i \triangleq \begin{cases} v_i, & |v_i| \geq \theta, \\ 0, & |v_i| < \theta, \quad i = 1, 2, \ldots, n, \end{cases} \tag{5.58}$$

with $\theta = \sqrt{2\gamma\lambda}$, that is,

$$\text{prox}_{\gamma f_2}(v) = H_{\sqrt{2\gamma\lambda}}(v). \tag{5.59}$$

See Exercise 4.21, p. 71 for details. As shown in Figure 5.6, the hard-thresholding operator rounds small elements ($|v_i| < \theta$) to 0. Figure 5.7 illustrates this operation. By using this operator, the proximal gradient algorithm for the ℓ^0 regularization (5.53) is given as follows.

IHT for ℓ^0 regularization (5.53)

Initialization: Give an initial vector $x[0]$ and positive number $\gamma > 0$.
Iteration: for $k = 0, 1, 2, \ldots$ do

$$x[k+1] = H_{\sqrt{2\gamma\lambda}}\big(x[k] - \gamma\,\Phi^\top(\Phi x[k] - y)\big). \tag{5.60}$$

This algorithm is called the *iterative hard-thresholding algorithm* (IHT).

---

3.   See Section 4.4 (p. 76).

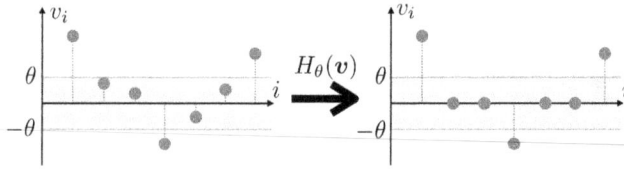

Figure 5.7. Hard-thresholding operator $H_\theta(v)$ rounds small elements ($|v_i| < \theta$) to 0, where $\theta = \sqrt{2\gamma\lambda}$.

For the convergence of the iterative hard-thresholding algorithm (5.60), the following theorem is proved in [9]:

Theorem 5.7. *Assume that*

$$\gamma < \frac{1}{\|\Phi\|^2} \tag{5.61}$$

holds where $\|\Phi\|$ is the maximum singular value of Φ. Then the sequence $\{x[0], x[1], x[2], \ldots\}$ generated by the iterative hard-thresholding algorithm (5.60) converges to a local minimizer of the ℓ^0 regularization (5.53). Moreover, the convergence is first order, that is, there exists a constant $c \in (0, 1)$ such that

$$\|x[k+1] - x^*\|_2 \le c\|x[k] - x^*\|_2, \quad k = 0, 1, 2, \ldots, \tag{5.62}$$

where x^* is a local minimizer.

The condition in (5.61) is very similar to the condition in (4.93) (p. 80) for the proximal gradient algorithm for ℓ^1 regularization. The convergence rate $O(c^k)$ of IHT is much faster than that of ISTA, $O(1/k)$ or FISTA, $O(1/k^2)$. Note however that IHT just gives a local minimizer, which is not necessarily equivalent to a global one.

5.3.2   Iterative *s*-sparse Algorithm

Here we consider the *s*-sparse approximation (5.54). By using the indicator function (4.37) in Chapter 4 (p. 68), we rewrite the constrained problem of *s*-sparse approximation (5.54) as an unconstrained optimization problem. Let us denote by Σ_s the set of *s*-sparse vectors in $\mathbb{R}^n$:

$$\Sigma_s \triangleq \{x \in \mathbb{R}^n : \|x\|_0 \le s\}. \tag{5.63}$$

Exercise 5.8. Show that Σ_s is a non-convex set.

The indicator function I_{Σ_s} for the set Σ_s is given by

$$I_{\Sigma_s}(x) = \begin{cases} 0, & \|x\|_0 \le s, \\ \infty, & \|x\|_0 > s. \end{cases} \tag{5.64}$$

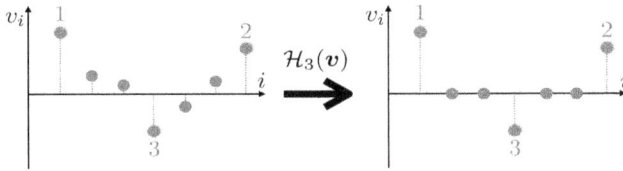

Figure 5.8. s-sparse operator $\mathcal{H}_s(v)$ with $s = 3$: the 3 largest elements in magnitude are unchanged and the other elements are set to 0. The numbers 1, 2, 3 indicates the rank of the absolute values of the elements.

By using this, the s-sparse approximation (5.54) is equivalently described by

$$\underset{x \in \mathbb{R}^n}{\text{minimize}} \quad \frac{1}{2}\|\Phi x - y\|_2^2 + I_{\Sigma_s}(x). \tag{5.65}$$

Note that since Σ_s is non-convex (see Exercise 5.8), the indicator function I_{Σ_s} is not a convex function. Anyhow, let us apply this to the proximal gradient algorithm (5.56). To do this, we should compute the proximal operator of the indicator function I_{Σ_s}, which is equal to the projection onto the set Σ_s. The projection is actually obtained by

$$\Pi_{\Sigma_s}(v) = \arg\min_{x \in \Sigma_s} \|x - v\|_2 = \mathcal{H}_s(v), \tag{5.66}$$

where $\mathcal{H}_s(v)$ is the s-sparse operator that sets all but the s largest (in magnitude) elements of v to 0. Figure 5.8 illustrates this operation. Note that the projection is in general not unique. If s largest elements are not uniquely determined, then they can be chosen either randomly or based on a fixed ordering rule.

Exercise 5.9. Prove that equation (5.66) holds.

Let $\gamma_s(v)$ denote the s-th largest elements of vector $v \in \mathbb{R}^n$. Then the s-sparse operator $\mathcal{H}_s(v)$ can be represented by using the hard-thresholding operator (5.58) as

$$\mathcal{H}_s(v) = H_{\gamma_s(v)}(v). \tag{5.67}$$

From this, the s-sparse operator is sometimes called as the hard-thresholding operator.

By using the s-sparse operator (5.66) as the proximal operator of the indicator function I_{Σ_s}, we obtain the proximal gradient algorithm (5.56) for the s-sparse approximation (5.65).

┌─ Iterative s-sparse algorithm for s-sparse approximation (5.54) ─────────

Initialization: Give an initial vector $x[0]$ and a positive number $\gamma > 0$
Iteration: for $k = 0, 1, 2, \ldots$ do

$$x[k+1] = \mathcal{H}_s\big(x[k] - \gamma\, \Phi^\top(\Phi x[k] - y)\big). \qquad (5.68)$$

We call this algorithm the *iterative s-sparse algorithm*. Note that this is sometimes called the *iterative hard-thresholding algorithm* (IHT).

For the iterative s-sparse algorithm, we have the following convergence theorem [9].

Theorem 5.10. *Assume that the matrix $\Phi \in \mathbb{R}^{m \times n}$ is surjective, that is,* $\mathrm{rank}(\Phi) = m$, *and the column vectors ϕ_i, $i = 1, 2, \ldots, n$, are non-zero, that is,*

$$\|\phi_i\|_2 > 0, \quad \forall i \in \{1, 2, \ldots, n\}. \qquad (5.69)$$

Assume also that the constant $\gamma > 0$ satisfies

$$\gamma < \frac{1}{\|\Phi\|^2}. \qquad (5.70)$$

Then the sequence $\{x[0], x[1], x[2], \ldots\}$ generated by the s-sparse algorithm (5.68) converges to a local minimizer of the s-sparse approximation problem (5.54). Moreover, the convergence is first order, that is, there exists a constant $c \in (0, 1)$ such that

$$\|x[k+1] - x^*\|_2 \le c\|x[k] - x^*\|_2, \quad k = 0, 1, 2, \ldots, \qquad (5.71)$$

where x^* is a local minimizer.

5.3.3   Compressive Sampling Matching Pursuit (CoSaMP)

For the s-sparse approximation (5.54), we can extend the algorithm of OMP in Section 5.2.2 with the s-sparse operator $\mathcal{H}_s$. This algorithm is called the *compressive sampling matching pursuit* (CoSaMP).

In the OMP algorithm (5.51), we first choose one index $i[k]$ as

$$i[k] = \arg\max_{i \in \{1, \ldots, n\}} \frac{\langle \phi_i, r[k-1]\rangle^2}{\|\phi_i\|_2^2}. \qquad (5.72)$$

Alternatively, CoSaMP chooses $2s$ largest values of

$$\frac{\langle \phi_i, r[k-1]\rangle^2}{\|\phi_i\|_2^2} = \left\langle \frac{\phi_i}{\|\phi_i\|_2}, r[k-1]\right\rangle^2, \qquad (5.73)$$

and includes these $2s$ indices in the index set $\mathcal{S}_k$, that is,

$$\mathcal{S}_k = \mathcal{S}_{k-1} \cup \mathrm{supp}\left\{ \mathcal{H}_{2s}\left(\left\langle \frac{\phi_i}{\|\phi_i\|_2}, r[k-1] \right\rangle^2 \right) \right\}. \tag{5.74}$$

As in OMP, we then find the projection of y onto the linear subspace $\mathcal{C}_k = \{\phi_i : i \in \mathcal{S}_k\}$. That is,

$$\tilde{x}[k] = \left(\Phi_{\mathcal{S}_k}^{\top} \Phi_{\mathcal{S}_k}\right)^{-1} \Phi_{\mathcal{S}_k}^{\top} y. \tag{5.75}$$

From this, we define an n-dimensional coefficient vector $z[k]$ as

$$(z[k])_i \triangleq \begin{cases} (\tilde{x}[k])_i, & i \in \mathcal{S}_k, \\ 0, & i \notin \mathcal{S}_k. \end{cases} \tag{5.76}$$

Note that the number of nonzero coefficients in $z[k]$ is larger than $2s$. We then *prune* $z[k]$ to an s-sparse vector $x[k]$ as

$$x[k] = \mathcal{H}_s\left(z[k]\right). \tag{5.77}$$

Also, we update the index set $\mathcal{S}_k$ to

$$\mathcal{S}_k = \mathrm{supp}(x[k]). \tag{5.78}$$

Finally, we obtain the CoSaMP algorithm to solve the s-sparse approximation (5.54).

CoSaMP algorithm for s-sparse approximation (5.54)

Initialization: Set $x[0] = 0$, $r[0] = y$, $\mathcal{S}_0 = \emptyset$
Iteration: for $k = 1, 2, \ldots$ do

$$\mathcal{I}[k] := \mathrm{supp}\left\{ \mathcal{H}_{2s}\left(\left\langle \frac{\phi_i}{\|\phi_i\|_2}, r[k-1] \right\rangle^2 \right) \right\},$$

$$\mathcal{S}_k := \mathcal{S}_{k-1} \cup \mathcal{I}[k],$$

$$\tilde{x}[k] := \left(\Phi_{\mathcal{S}_k}^{\top} \Phi_{\mathcal{S}_k}\right)^{-1} \Phi_{\mathcal{S}_k}^{\top} y,$$

$$(z[k])_{\mathcal{S}_k} := \tilde{x}[k], \tag{5.79}$$

$$(z[k])_{\mathcal{S}_k^c} := 0,$$

$$x[k] := \mathcal{H}_s\left(z[k]\right),$$

$$\mathcal{S}_k := \mathrm{supp}\{x[k]\},$$

$$r[k] := y - \Phi_{\mathcal{S}_k}\tilde{x}[k].$$

For the convergence of the CoSaMP algorithm, see the original paper [88].

5.4   Numerical Example

Here we solve sparse optimization numerically by using the greedy algorithms studied in this chapter. Let us consider the problem of curve fitting studied in Section 3.3 (p. 50). with the sparse polynomial $y = -t^{80} + t$. The data are given as in Section 3.3, the data points are given by

$$t_i = 0.1(i - 1), \quad i = 1, 2, \ldots, 11, \tag{5.80}$$

from which we reconstruct the 80-th order polynomial. Here we consider the following 6 algorithms:

1. ℓ^1 optimization considered in Section 3.3 (p. 50)
2. matching pursuit (MP)
3. orthogonal matching pursuit (OMP)
4. iterative hard-thresholding (IHT)
5. iterative s-sparse algorithm (ISS)
6. compressive sampling matching pursuit (CoSaMP)

The matrix $\Phi \in \mathbb{R}^{11 \times 81}$ is given by (3.16), which satisfies

$$0.012 < \frac{1}{\|\Phi\|^2} < 0.013, \tag{5.81}$$

and we choose the parameter γ for IHT and ISS as

$$\gamma = 0.01 < 0.012 < \frac{1}{\|\Phi\|^2}. \tag{5.82}$$

From Theorems 5.7 and 5.10, the condition (5.82) guarantees convergence to local minimizers for IHT and ISS. We also choose λ in the ℓ^0 regularization problem as $\lambda = 0.001$.

Figure 5.9 shows the coefficients obtained by the algorithms. The coefficients are ordered from the highest degree to the lowest degree. We see that the ℓ^1 optimization, MP, OMP, and CoSaMP give exact coefficients, while IHT and ISS show incorrect reconstruction. To see this more precisely, we check the estimation error $r = y - \Phi x^*$, where x^* is the obtained vector when the algorithm stops. Table 5.1 shows the error with the number of iterations required to achieve the error.

All but ℓ^1 optimization stop the iteration when the error $\|r[k]\|_2$ is less than 10^{-5} or the number of iterations is larger than 10^5.

IHT and ISS attained the maximum number of iterations 10^5, and their errors are much larger than those of the other methods. This is because they were trapped into local minimizers. The other methods show fast convergence, among which

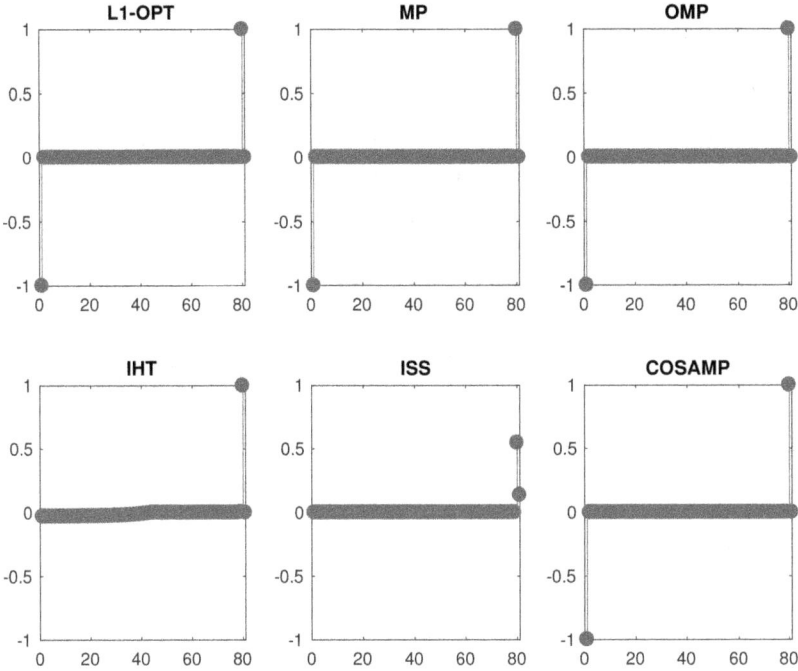

Figure 5.9. Estimation of sparse coefficients.

Table 5.1. Estimation error $\|y - \Phi x^*\|_2$ and number of iterations. IHT and ISS reached the maximum number 10^5 of iterations.

Methods	ℓ^1 OPT	MP	OMP	IHT	ISS	CoSaMP
Error	2.7×10^{-10}	9.1×10^{-6}	4.1×10^{-16}	0.0017	0.83	4.1×10^{-11}
Iterations	10	18	2	10^5	10^5	3

OMP (2 iterations) and CoSaMP (3 iterations) especially present surprising results. In view of the error and the number of iterations, OMP is the best method in this case. It should be noted that greedy algorithms do not necessarily give a global solution. OMP is the best in this case but in other cases, another method may be the best. This depends on the problem and data, and we should adopt trial and error to seek the best algorithm.

5.5   Further Reading

Basics of greedy algorithms can be found in [26, 62]. For the characterization of ℓ^0
optimality by using the mutual coherence, or the *restricted isometry property* (RIP),
which is not studied in this book, you can refer to [37, 38].

The matching pursuit (MP) was first proposed in [71], the orthogonal matching
pursuit in [28, 94]. For the iterative hard-thresholding algorithm and the iterative
s-sparse algorithm, see the paper [9]. The compressive sampling matching pursuit
(CoSaMP) was proposed in [88].

MATLAB function of MP

MP.m

```
function [x,nitr]=MP(y,Phi,EPS,MAX_ITER)
  [m,n] = size(Phi);
  x = zeros(n,1);
  r = y;
  k = 0;
  Phi_norm = diag(Phi'*Phi);
  while (norm(r)>EPS) & (k < MAX_ITER)
    p = Phi'*r;
    v = p./sqrt(Phi_norm);
    [z,ik] = max(abs(v));
    v2 = p./Phi_norm;
    z = v2(ik);
    x(ik) = x(ik)+z;
    r = r-z*Phi(:,ik);
    k = k+1;
  end
  nitr=k;
end
```

MATLAB function of OMP

OMP.m

```
function [x,nitr]=OMP(y,Phi,EPS,MAX_ITER)
    [m,n] = size(Phi);
    x = zeros(n,1);
    r = y;
    k = 0;
    S = zeros(n,1);
    Phi_norm = diag(Phi'*Phi);
    while  (norm(r)>EPS) & (k < MAX_ITER)
        p = Phi'*r;
        v = p./sqrt(Phi_norm);
        [z,ik] = max(abs(v));
        S(ik) = ik;
        Phi_S = Phi(:,S(S>0));
        x(S(S>0)) = pinv(Phi_S)*y;
        r = y-Phi*x;
        k = k+1;
    end
    nitr=k;
end
```

MATLAB function of hard-thresholding operator $H_\lambda(v)$

hard_thresholding.m

```
function hv = hard_thresholding(lambda,v)
    [m,n]=size(v);
    mn = m*n;
    hv = zeros(m,n);
    for i = 1:mn
        if abs(v(i))<=lambda
            hv(i) = 0;
        else
            hv(i) = v(i);
        end
    end
end
```

MATLAB function of the support of a vector

supp.m

```matlab
function I = supp(x)
    I = find(abs(x)>0)';
end
```

MATLAB function of iterative hard-thresholding algorithm

IHT.m

```matlab
function [x,nitr]=IHT(y,Phi,lambda,gamma,EPS,MAX_ITER)
    [m,n] = size(Phi);
    x = zeros(n,1);
```

```matlab
    r = y;
    k = 0;
    while (norm(r)>EPS) & (k < MAX_ITER)
        p = x + gamma * Phi'*r;
        x = hard_thresholding(sqrt(2*lambda*gamma),p);
        S = supp(x);
        r = y-Phi(:,S)*x(S);
        k = k+1;
    end
    nitr=k;
end
```

MATLAB function of *s*-sparse operator

s_sparse_operator.m

```matlab
function y = s_sparse_operator(x,s)
   [n,m]=size(x);
   y=zeros(n,m);
   [xs,indx]=sort(abs(x),1,'descend');
   indx_s = indx(1:s);
   y(indx_s)=x(indx_s);
end
```

MATLAB function of iterative *s*-sparse algorithm

iterative_s_sparse.m

```matlab
function [x,nitr]=iterative_s_sparse(y,Phi,s,gamma,EPS,MAX_ITER)
   [m,n] = size(Phi);
   x = zeros(n,1);
   r = y;
   k = 0;
   while (norm(r)>EPS) & (k < MAX_ITER)
      p = x + gamma * Phi'*r;
      x = s_sparse_operator(p,s);
      S = supp(x);
      r = y-Phi(:,S)*x(S);
      k = k+1;
   end
   nitr=k;
end
```

MATLAB function of CoSaMP

CoSaMP.m

```matlab
function [x,nitr]=CoSaMP(y,Phi,s,EPS,MAX_ITER)
  [m,n] = size(Phi);
  x = zeros(n,1);
  r = y;
  k = 0;
  S = [];
  Lambda = [];
  Phi_norm = diag(Phi'*Phi);
  while (norm(r)>EPS) & (k < MAX_ITER)
    p = s_sparse_operator((Phi'*r)./sqrt(Phi_norm),2*s);
    Ik = supp(p);
    S = union(Lambda,Ik);
    Phi_S = Phi(:,S);
    z = zeros(n,1);
    z(S) = pinv(Phi_S)*y;
    x = s_sparse_operator(z,s);
    Lambda = supp(x);
    r = y-Phi_S*z(S);
    k = k+1;
  end
  nitr=k;
end
```

MATLAB code for the simulation in Section 5.4

```matlab
clear;
%% data
% polynomial coefficients
x_orig = [-1,zeros(1,78),1,0]';
% sampling
t = 0:0.1:1;
y = polyval(x_orig,t)';
% data size
N = length(t);
M = N-1;
% Order of polynomial
M_l = length(x_orig)-1;
% Vandermonde matrix
Phi=[];
for m=0:M_l
  Phi = [t'.^m,Phi];
end
```

```
%% Sparse modeling
% iteration parameters
EPS=1e-5; % if the residual < EPS then the iteration will stop
MAX_ITER=100000; % maximum number of iterations
% L1 by CVX
cvx_begin
    variable x_l1(M_l+1)
    minimize norm(x_l1,1)
    subject to
        Phi*x_l1 == y
cvx_end
% Matching Pursuit
[x_mp,nitr_mp]=MP(y,Phi,EPS,MAX_ITER);
% OMP
[x_omp,nitr_omp]=OMP(y,Phi,EPS,MAX_ITER);
% CoSaMP
s = length(supp(x_orig));
[x_cosamp,nitr_cosamp]=CoSaMP(y,Phi,s,EPS,MAX_ITER);
% IHT
lambda=0.001;
gamma=0.01;
[x_iht,nitr_iht]=IHT(y,Phi,lambda,gamma,EPS,MAX_ITER);
% iterative s-sparse
gamma=0.01;
[x_iss,nitr_iss]=iterative_s_sparse(y,Phi,s,gamma,EPS,MAX_ITER);
```

DOI: 10.1561/9781680837254.ch6

Chapter 6

Applications of Sparse Representation

In this section, we introduce applications of sparse representation in a finite-dimensional space for systems and control.

6.1 Sparse Representations for Splines

In this section, we consider curve fitting by using splines. As discussed in Chapter 3, we consider the following two-dimensional data:

$$\mathcal{D} = \{(t_1, y_1), (t_2, y_2), \ldots, (t_m, y_m)\}, \tag{6.1}$$

where $0 \leq t_1 < t_2 < \cdots < t_m = T$ are sampling times, and $y_1, y_2, \ldots, y_m$ are obtained from the following observation:

$$y_i = y(t_i) + \epsilon_i, \quad i = 1, 2, \ldots, m, \tag{6.2}$$

where y is a function and ϵ_i is additive noise. The problem of curve fitting is to estimate the unknown function y form data $\mathcal{D}$.

In Chapter 3, we have assumed the function is a polynomial function, and shown that with a fixed order of the polynomial, the problem becomes a convex optimization. Here we seek a function among more general functions called *splines*. Namely, we consider the following optimization problem:

$$\underset{y}{\text{minimize}} \sum_{i=1}^{m} |y(t_i) - y_i|^2 + \lambda \int_0^T |\ddot{y}(t)|^2 dt, \tag{6.3}$$

where we assume the second derivative $\ddot{y}$ is in $L^2(0, T)$. The first term is for the fidelity of curve fitting to the data, and the second term is for the smoothness of the curve. In general, if you increase the fidelity then the curve becomes less smooth, and hence we need to control the trade-off between them to appropriately choose the parameter $\lambda > 0$.

Note that since y is not a finite-dimensional vector but a function, the problem is an *infinite-dimensional problem*. However, to use techniques in Hilbert space theory, the problem can be reduced to a finite-dimensional optimization problem. Let us first show this in this section. For this, we introduce the formulation of *control theoretic splines* [36, 108].

6.1.1   Solution by Projection Theorem

First, let us define

$$x_1(t) \triangleq y(t), \quad x_2(t) \triangleq \dot{y}(t), \quad u(t) \triangleq \ddot{y}(t). \tag{6.4}$$

Then, the optimization problem can be described as

$$\underset{u \in L^2(0,T)}{\text{minimize}} \sum_{i=1}^{m} |y(t_i) - y_i|^2 + \lambda \int_0^T |u(t)|^2 dt$$

$$\text{subject to} \quad \dot{x}(t) = Ax(t) + bu(t), \quad y(t) = c^\top x(t), \quad t \in [0, T] \tag{6.5}$$

$$x(0) = 0$$

where $x(t) \triangleq [x_1(t), x_2(t)]^\top$ and

$$A \triangleq \begin{bmatrix} 0 & 1 \\ 0 & 0 \end{bmatrix}, \quad b \triangleq \begin{bmatrix} 0 \\ 1 \end{bmatrix}, \quad c \triangleq \begin{bmatrix} 1 \\ 0 \end{bmatrix}. \tag{6.6}$$

Note that this formulation is for more general optimization than (6.3) by choosing another set of (A, b, c).

Define

$$l(\tau, t) \triangleq \begin{cases} c^\top e^{A(t-\tau)} b, & \text{if } 0 \leq t \leq \tau \\ 0, & \text{otherwise} \end{cases} \tag{6.7}$$

and

$$\phi_i(t) \triangleq l(t, t_i), \quad i = 1, 2, \dots, m. \tag{6.8}$$

Then we have

$$y(t_i) = \langle \phi_i, u \rangle_{L^2} = \int_0^T \phi_i(t)u(t)dt, \quad i = 1, 2, \dots, m. \tag{6.9}$$

From this, the problem (6.5) becomes

$$\underset{u \in L^2(0,T)}{\text{minimize}} \sum_{i=1}^m \left| \langle \phi_i, u \rangle_{L^2} - y_i \right|^2 + \lambda \int_0^T |u(t)|^2 dt. \tag{6.10}$$

Then, if we define $z_i \triangleq \langle \phi_i, u \rangle_{L^2}$, the optimization problem is described as

$$\underset{u \in L^2(0,T)}{\text{minimize}} \sum_{i=1}^m |z_i - y_i|^2 + \lambda \int_0^T |u(t)|^2 dt \tag{6.11}$$

$$\text{subject to} \quad z_i = \langle \phi_i, u \rangle_{L^2}, \quad i = 1, 2, \dots, m.$$

Define a new Hilbert space H by

$$H = L^2(0, T) \times \mathbb{R}^m, \tag{6.12}$$

with inner product

$$\left\langle \begin{bmatrix} u \\ z \end{bmatrix}, \begin{bmatrix} v \\ w \end{bmatrix} \right\rangle_H \triangleq w^\top z + \int_0^T u(t)v(t)dt. \tag{6.13}$$

Then, consider a closed linear subspace M of H defined by

$$M \triangleq \left\{ \begin{bmatrix} u \\ z \end{bmatrix} \in H : z_i = \langle \phi_i, u \rangle_{L^2} \right\}, \tag{6.14}$$

and a vector $p \in H$ defined by

$$p \triangleq \begin{bmatrix} 0 \\ y \end{bmatrix} \in H, \tag{6.15}$$

where $y = [y_1, y_2, \dots, y_m]^\top \in \mathbb{R}^m$. Then, for $r \triangleq (u, z) \in H$, we have

$$\|r - p\|_H^2 = \sum_{i=1}^m |z_i - y_i|^2 + \lambda \int_0^T |u(t)|^2 dt, \tag{6.16}$$

where $\| \cdot \|_H$ is the norm induced by the inner product $\langle \cdot, \cdot \rangle_H$, that is

$$\|r - p\|_H = \sqrt{\langle r - p, r - p \rangle_H}. \tag{6.17}$$

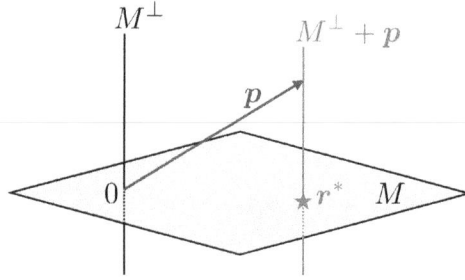

Figure 6.1. Projection theorem: the projection of p onto M is given by $r^* \in (M^\perp + p) \cap M$.

The optimization problem (6.11) is now rewritten as

$$\underset{r \in H}{\text{minimize}} \, \|r - p\|_H^2 \quad \text{subject to} \quad r \in M. \tag{6.18}$$

The minimizer is given by the *projection* of $p \in H$ onto the closed linear subspace $M \subset H$. Let $M^\perp$ denote the *orthogonal complement* of M in H. That is,

$$M^\perp \triangleq \left\{ \begin{bmatrix} v \\ w \end{bmatrix} : \left\langle \begin{bmatrix} v \\ w \end{bmatrix}, \begin{bmatrix} u \\ z \end{bmatrix} \right\rangle_H = 0, \, \forall \begin{bmatrix} u \\ z \end{bmatrix} \in M \right\}. \tag{6.19}$$

Then, from the projection theorem, the minimizer r^* is in the set $(M^\perp + p) \cap M$ (see [36, Section 2.3]). Figure 6.1 illustrates this property form the projection theorem.

Now, let us characterize the set $M^\perp + p$. Take $(v, w) \in M^\perp$. Then, from (6.14), for any $(u, z) \in M$, we have

$$
\begin{aligned}
0 &= \left\langle \begin{bmatrix} v \\ w \end{bmatrix}, \begin{bmatrix} u \\ z \end{bmatrix} \right\rangle_H \\
&= w^\top z + \lambda \int_0^T v(t) u(t) dt \\
&= w^\top \begin{bmatrix} \langle \phi_1, u \rangle_{L^2} \\ \langle \phi_2, u \rangle_{L^2} \\ \vdots \\ \langle \phi_m, u \rangle_{L^2} \end{bmatrix} + \lambda \langle v, u \rangle_{L^2} \\
&= \sum_{i=1}^m w_i \langle \phi_i, u \rangle_{L^2} + \lambda \langle v, u \rangle_{L^2} \\
&= \left\langle \sum_{i=1}^m w_i \phi_i + \lambda v, u \right\rangle_{L^2}
\end{aligned}
\tag{6.20}
$$

This equation holds for any $u \in L^2(0, T)$, and hence we have

$$\sum_{i=1}^{m} w_i \phi_i + \lambda v = 0, \tag{6.21}$$

or

$$v = -\frac{1}{\lambda} \sum_{i=1}^{m} w_i \phi_i. \tag{6.22}$$

From this, the subspace $M^\perp$ can be represented by

$$M^\perp = \left\{ \begin{bmatrix} -\frac{1}{\lambda} \sum_{i=1}^{m} w_i \phi_i \\ w \end{bmatrix} : w \in \mathbb{R}^m \right\}, \tag{6.23}$$

and also we have

$$M^\perp + p = \left\{ \begin{bmatrix} -\frac{1}{\lambda} \sum_{i=1}^{m} w_i \phi_i \\ w + y \end{bmatrix} : w \in \mathbb{R}^m \right\}. \tag{6.24}$$

Now, let us obtain the minimizer $r^* = (u^*, z^*) \in (M^\perp + p) \cap M$ of (6.18) (see also Figure 6.1). First, since $r^* \in M$, we have

$$z_i^* = \langle \phi_i, u^* \rangle_{L^2}, \quad i = 1, 2, \ldots, m. \tag{6.25}$$

Next, since $r^* \in M^\perp + p$, we have

$$u^* = -\frac{1}{\lambda} \sum_{i=1}^{m} w_i \phi_i \tag{6.26}$$

$$z_i^* = w_i + y_i \tag{6.27}$$

Inserting (6.26) into (6.25) gives

$$z_i^* = \left\langle \phi_i, -\frac{1}{\lambda} \sum_{j=1}^{m} w_j \phi_j \right\rangle_{L^2} = -\frac{1}{\lambda} \sum_{j=1}^{m} w_j \langle \phi_i, \phi_j \rangle_{L^2}. \tag{6.28}$$

From (6.27), we have

$$-\frac{1}{\lambda} \sum_{j=1}^{m} w_j \langle \phi_i, \phi_j \rangle_{L^2} = w_i + y_i, \tag{6.29}$$

or

$$(\lambda I + G)w = -\lambda y, \tag{6.30}$$

where G is the Gram matrix defined by

$$G \triangleq \begin{bmatrix} \langle \phi_1, \phi_1 \rangle & \langle \phi_1, \phi_2 \rangle & \cdots & \langle \phi_1, \phi_m \rangle \\ \langle \phi_2, \phi_1 \rangle & \langle \phi_2, \phi_2 \rangle & \cdots & \langle \phi_2, \phi_m \rangle \\ \vdots & \vdots & \ddots & \vdots \\ \langle \phi_m, \phi_1 \rangle & \langle \phi_m, \phi_2 \rangle & \cdots & \langle \phi_m, \phi_m \rangle \end{bmatrix}. \tag{6.31}$$

Since $\lambda > 0$, the matrix $\lambda I + G$ is non-singular, and hence

$$w = -\lambda(\lambda I + G)^{-1}y. \tag{6.32}$$

Finally, from (6.26), we obtain the solution

$$u^* = -\frac{1}{\lambda} \sum_{i=1}^{m} [-(\lambda I + G)^{-1}\lambda y]_i \phi_i$$

$$= \sum_{i=1}^{m} [(\lambda I + G)^{-1}y]_i \phi_i \tag{6.33}$$

$$= \sum_{i=1}^{m} \alpha_i^* \phi_i$$

where

$$\alpha^* = \begin{bmatrix} \alpha_1^* \\ \vdots \\ \alpha_m^* \end{bmatrix} = (\lambda I + G)^{-1}y. \tag{6.34}$$

The important point of the solution is that the optimal solution of the infinite-dimensional optimization problem in (6.5) is describe as a finite number of *spline functions* $\phi_1, \ldots, \phi_m$ and the problem is reduced to computing the unknown coefficients $\alpha_1^*, \ldots, \alpha_m^*$. In other words, the original problem (6.5) is fundamentally a finite-dimensional optimization problem. Note that this property is generalized to the *representer theorem* in statistical machine learning [104].

Finally, the optimal solution y^* of the original optimization problem (6.3) is given by

$$y^*(t) = \int_0^t \int_0^\tau u^*(s)ds d\tau. \tag{6.35}$$

6.1.2 Sparse Representation

From (6.33), the number of coefficients is equal to m, the number of data. If the data is very big (i.e., m is very large), then we need many coefficients to represent the fitting curve $y(t)$. Then, to use the idea of sparse representation, we can reduce the number of coefficients. For this, we restrict the feasible solutions of the optimization problem (6.10) to be

$$u(t) = \sum_{i=1}^{m} z_i \phi_i(t), \tag{6.36}$$

where $z_1, \ldots, z_m$ are unknown coefficients to be obtained. With this, we have

$$\langle \phi_i, u \rangle_{L^2} = \left\langle \phi_i, \sum_{j=1}^{m} z_j \phi_j \right\rangle_{L^2} = \sum_{j=1}^{m} z_j \langle \phi_i, \phi_j \rangle_{L^2}, \tag{6.37}$$

and hence

$$\begin{bmatrix} \langle \phi_1, u \rangle_{L^2} \\ \langle \phi_2, u \rangle_{L^2} \\ \vdots \\ \langle \phi_m, u \rangle_{L^2} \end{bmatrix} = Gz, \tag{6.38}$$

and

$$\sum_{i=1}^{m} \left| \langle \phi_i, u \rangle_{L^2} - y_i \right| = \| Gz - y \|^2. \tag{6.39}$$

Also, we have

$$\lambda \int_0^T |u(t)|^2 dt = \lambda \int_0^T \left(\sum_{i=1}^{m} z_i \phi_i(t) \right) \left(\sum_{j=1}^{m} z_j \phi_j(t) \right) dt$$

$$= \lambda \sum_{i=1}^{m} \sum_{j=1}^{m} z_i z_j \langle \phi_i, \phi_j \rangle_{L^2} \tag{6.40}$$

$$= \lambda z^\top G z.$$

Therefore, under the assumption of (6.36), the optimization problem (6.10) is rewritten as

$$\underset{z \in \mathbb{R}^m}{\text{minimize}} \, \| Gz - y \|^2 + \lambda z^\top G z. \tag{6.41}$$

Then, to promote the sparsity of z, we add the ℓ^0 norm as a regularization term:

$$\underset{z \in \mathbb{R}^m}{\text{minimize}} \, \|Gz - y\|^2 + \lambda z^\top Gz + \rho \|z\|_0, \qquad (6.42)$$

where $\rho > 0$ is the regularization parameter. As usual, we can adopt the ℓ^1 norm as convex relaxation of the ℓ^0 norm. The relaxed convex optimization problem is described as follows:

$$\underset{z \in \mathbb{R}^m}{\text{minimize}} \, \|Gz - y\|^2 + \lambda z^\top Gz + \rho \|z\|_1. \qquad (6.43)$$

This can be easily solved by the proximal gradient algorithm studied in Section 4.4 (p. 76).

6.2 Discrete-time Hands-off Control

In this section, we introduce sparse control (or hands-off control) for discrete-time systems.

6.2.1 Feasible Control

Let us consider a linear time-invariant system described by

$$x[k+1] = Ax[k] + bu[k], \quad k = 0, 1, 2, \ldots, n-1, \qquad (6.44)$$

where $x[k] \in \mathbb{R}^d$ is the state and $u[k] \in \mathbb{R}$ is the control at time step $k \in \{0, 1, 2, \ldots, n-1\}$. The matrix $A \in \mathbb{R}^{d \times d}$ and the vector $b \in \mathbb{R}^d$ are assumed to be exactly known. The number n is the *horizon length* of the system. We call the difference equation (6.44) the *state equation*.

Assume that the initial state $x[0] = \xi$ is given by state observation. Then the control objective is to find a control sequence $\{u[0], u[1], \ldots, u[n-1]\}$ such that the control drives the state $x[k]$ from $x[0] = \xi$ to the origin, that is,

$$x[n] = 0. \qquad (6.45)$$

From the state equation (6.44), we have

$$x[k] = A^k x[0] + \sum_{i=0}^{k-1} A^{k-1-i} bu[i] = A^k \xi + \sum_{i=0}^{k-1} A^{k-1-i} bu[i], \qquad (6.46)$$

for $k \in \{0, 1, \ldots, n-1\}$. Then, the terminal constraint (6.45) can be described as

$$x[n] = A^n \xi + \sum_{i=0}^{n-1} A^{n-1-i} bu[i] = A^n \xi + \Phi u = 0, \qquad (6.47)$$

where

$$\Phi \triangleq \begin{bmatrix} A^{n-1}b & A^{n-2}b & \cdots & Ab & b \end{bmatrix}, \quad u \triangleq \begin{bmatrix} u[0] \\ u[1] \\ \vdots \\ u[n-1] \end{bmatrix} \qquad (6.48)$$

Then the set of feasible controls that achieve (6.45) is given by

$$\mathcal{U}(n, \xi) \triangleq \{u \in \mathbb{R}^n : A^n \xi + \Phi u = 0\}. \qquad (6.49)$$

For the feasibility, we have the following lemma.

Lemma 6.1. *Suppose $n \geq d$ and the following matrix M is nonsingular:[1]*

$$M \triangleq \begin{bmatrix} b & Ab & \cdots & A^{d-1}b \end{bmatrix} \in \mathbb{R}^{d \times d}. \qquad (6.50)$$

Then the feasible set $\mathcal{U}(n, \xi)$ is non-empty for any $\xi \in \mathbb{R}^d$.

Proof: Since $n \geq d$ and the matrix M is non-singular, the matrix Φ in (6.48) has full row rank. It follows that Φ is surjective and there exists at least one vector u that satisfies $\Phi u = -A^n \xi$ for any $\xi \in \mathbb{R}^d$. □

6.2.2 Maximum Hands-off Control

The optimal control is a problem to seek the optimal solution(s) that minimizes a cost function among control vectors in $\mathcal{U}(n, \xi)$. A general form of the cost function is given by

$$J(u) = \sum_{k=0}^{n-1} L(x[k], u[k]), \qquad (6.51)$$

where the function L is called the *stage cost function*.

The *linear quadratic control*, or *LQ control* for short, has the following stage cost function

$$L(x, u) = x^\top Q x + r|u|^2, \qquad (6.52)$$

---

1.   The matrix M is called the *controllability matrix*, and the pair (A, b) is called *controllable* if M is non-singular.

where $Q \in \mathbb{R}^{d \times d}$ is a positive semidefinite matrix, and $r > 0$. In this section, we are interested in *sparse control*, also known as *maximum hands-off control*, which has the following stage cost function:

$$L(x, u) = |u|^0, \tag{6.53}$$

with which the cost function is given by

$$J(u) = \sum_{k=0}^{n-1} |u[k]|^0 = \|u\|_0. \tag{6.54}$$

The optimization problem is then described as

$$\operatorname*{minimize}_{u \in \mathbb{R}^n} \|u\|_0 \text{ subject to } u \in \mathcal{U}(n, \xi). \tag{6.55}$$

As usual, we approximate the ℓ^0 optimization by

$$\operatorname*{minimize}_{u \in \mathbb{R}^n} \|u\|_1 \text{ subject to } u \in \mathcal{U}(n, \xi), \tag{6.56}$$

which is the ℓ^1 optimization problem discussed in Section 4.3, which is efficiently solved by Douglas-Rachford splitting algorithm (see Section 4.3.1).

Also one can consider the following cost function

$$J(u) = \sum_{k=0}^{n-1} \left\{ x[k]^\top Q x[k] + \lambda |u[k]| \right\}, \tag{6.57}$$

with positive semidefinite $Q \in \mathbb{R}^{d \times d}$ and $\lambda > 0$. Inserting (6.46) into (6.57), we have

$$J(u) = u^\top R u + 2q^\top u + \lambda \|u\|_1 + c, \tag{6.58}$$

for some $R \in \mathbb{R}^{n \times n}$, $q \in \mathbb{R}^n$, and $c \in \mathbb{R}$. For this optimization, we can apply the ADMM algorithm discussed in Section 4.5.

6.2.3   Model Predictive Control

As discussed above, the control sequence $u \in \mathbb{R}^n$ is obtained by numerical optimization with given initial state observation $\xi \in \mathbb{R}^d$. Let denote by $\mathcal{C}$ the mapping from the initial state $\xi \in \mathbb{R}^d$ to the optimal control sequence $u \in \mathbb{R}^n$, that is,

$$u = \mathcal{C}(\xi). \tag{6.59}$$

Then $u = C(\xi)$ is a finite-horizon control (i.e., the control is applied to a plant in a finite length of time), and this is *open-loop control*. Open-loop control is something like riding a bicycle with your eyes closed, which is very fragile against disturbances. To make the control system *robust*, you need to implement the control as *feedback control*, where the controller constantly observes the state and update the control based on the latest state observation.

To implement feedback control from the finite-horizon control $u = C(\xi)$, we adopt the *model predictive control* (also known as *receding horizon control*).

The model predictive control is described as follows:

1. Observe the state $x[k]$ at time k.
2. Compute the optimal control sequence

$$u[k] = \begin{bmatrix} u_0[k] \\ u_1[k] \\ \vdots \\ u_{n-1}[k] \end{bmatrix} = C(x[k]). \tag{6.60}$$

3. Use the first element of $u[k]$, that is, $u_0[k]$, as the control at time k.

From this, the control $u[k]$ to the discrete-time plant (6.44) is obtained by

$$u[k] = u_0[k] = \begin{bmatrix} 1 & 0 & \cdots & 0 \end{bmatrix} C(x[k]). \tag{6.61}$$

Figure 6.2 shows the block diagram of the feedback control system where P is the plant.

The important thing we should do next is to study the *stability* of the feedback system. The closed-loop system in Figure 6.2 may exhibit instability, that is, $x[k]$ may diverge, if we do not care about the stability. The instability is possible even when P and C are both stable. Therefore, to prove the stability is very important to design a feedback control system.

First, we define the *value function* $V(\xi)$ of the optimal control problem (6.56) by

$$V(\xi) \triangleq \min_{u \in \mathcal{U}(n,\xi)} \|u\|_1. \tag{6.62}$$

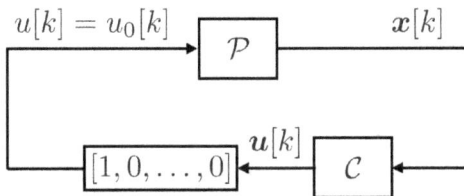

Figure 6.2. Feedback control by model predictive control $u[k] = C(x[k])$. P is the plant given by (6.44).

We have the following lemma:

Lemma 6.2. *Assume that the controllability matrix M in (6.50) and the matrix A in (6.44) are non-singular. Assume also that n ≥ d. Then the value function V (ξ) is convex, continuous, and positive definite.*

Exercise 6.3. *Prove Lemma 6.2.*

Now, we give a detailed definition of stability.

Definition 6.4. *Let us consider the following discrete-time system*

$$x[k + 1] = f(x[k]), \quad k = 0, 1, 2, \ldots \tag{6.63}$$

Suppose that there exists the unique sequence $\{x[0], x[1], \ldots\}$ satisfying (6.63) for any initial state $x[0] \in \mathbb{R}^d$. Suppose also that the origin is an equilibrium of the system, namely, $f(0) = 0$ holds. Then the origin is said to be stable *if for each $\epsilon > 0$ there exists $\delta > 0$ such that*

$$\|x[0]\|_2 < \delta \Rightarrow \|x[k]\|_2 < \epsilon, \quad \forall k \geq 0. \tag{6.64}$$

The concept is very simple; the state trajectory $\{x[k]\}_{k=0}^{\infty}$ starting out near the origin will keep on staying near the origin and never diverge.

From (6.44) and (6.61), the closed-loop system is described as

$$x[k + 1] = Ax[k] + \begin{bmatrix} 1 & 0 & \ldots & 0 \end{bmatrix} C(x[k]) \triangleq f(x[k]). \tag{6.65}$$

It is easily shown that the origin **0** is an equilibrium of this difference equation. To show the stability of this equilibrium, *Lyapunov's theorem* is available.

Theorem 6.5. *Suppose that there exists a function $V : \mathbb{R}^d \to \mathbb{R}$ satisfying*

1. *$V(0) = 0$.*
2. *$V(ξ)$ is continuous.*
3. *$V(ξ) > 0$ for any $ξ \neq 0$.*
4. *$V(x[k+1]) \leq V(x[k])$ for $k = 0, 1, 2, \ldots$, for the state trajectory $\{x[k]\}_{k=0}^{\infty}$ of the system (6.63).*

*Then the origin **0** is stable under the system equation (6.63).*

A function V in Theorem 6.5 is called a *Lyapunov function*. The idea to prove the stability of our system (6.65) is to show the value function (6.62) to be a Lyapunov function. In fact, it is a Lyapunov function and we have the following theorem.

Theorem 6.6. *Assume M and A are non-singular, and n ≥ d. Then the origin is stable under the system equation (6.65).*

Proof: We prove the value function $V(\xi)$ in (6.62) is a Lyapunov function of (6.65). The properties 1 to 3 in Theorem 6.5 are directly from Lemma 6.2. We here prove 4. Let

$$\boldsymbol{u}^*[k] \triangleq \begin{bmatrix} u_0^*[k] & u_1^*[k] & \cdots & u_{n-1}^*[k] \end{bmatrix}^\top = \mathcal{C}(\boldsymbol{x}[k]), \qquad (6.66)$$

and define

$$\tilde{\boldsymbol{u}}[k] \triangleq \begin{bmatrix} u_1^*[k] & u_2^*[k] & \cdots & u_{n-1}^*[k], 0 \end{bmatrix}^\top. \qquad (6.67)$$

Note that $\tilde{\boldsymbol{u}}[k]$ is a shifted control sequence by one time step of the optimal control sequence $\boldsymbol{u}^*[k]$ at time k. It is then easily shown that $\tilde{\boldsymbol{u}}[k]$ is a feasible control for $\boldsymbol{x}[k+1]$, that is,

$$\tilde{\boldsymbol{u}}[k] \in \mathcal{U}(n, \boldsymbol{x}[k+1]). \qquad (6.68)$$

In fact, since $\boldsymbol{u}^*[k] \in \mathcal{U}(n, \boldsymbol{x}[k])$ we have

$$
\begin{aligned}
A^n \boldsymbol{x}[k+1] &+ \Phi \tilde{\boldsymbol{u}}[k] \\
&= A^n (A\boldsymbol{x}[k] + bu_0^*[k]) + A^{n-1} bu_1^*[k] + \cdots + Abu_{n-1}^*[k] \\
&= A\big(A^n \boldsymbol{x}[k] + A^{n-1} bu_0^*[k] + A^{n-2} bu_1^*[k] + \cdots + bu_{n-1}^*[k]\big) \\
&= A\big(A^n \boldsymbol{x}[k] + \Phi \boldsymbol{u}^*[k]\big) \\
&= A \times \mathbf{0} \\
&= \mathbf{0}.
\end{aligned}
\qquad (6.69)
$$

Then, from the optimality of the value function, we have

$$
\begin{aligned}
V(\boldsymbol{x}[k+1]) &= \min\{\|\boldsymbol{u}\|_1 : \boldsymbol{u} \in \mathcal{U}(n, \boldsymbol{x}[k+1])\} \\
&\leq \|\tilde{\boldsymbol{u}}[k]\|_1 \\
&= |u_1^*[k]| + |u_2^*[k]| + \cdots + |u_{n-1}^*[k]| + |0| \\
&= \sum_{i=0}^{n-1} |u_i^*[k]| - |u_0^*[k]| \\
&= V(\boldsymbol{x}[k]) - |u_0^*[k]| \\
&\leq V(\boldsymbol{x}[k]),
\end{aligned}
\qquad (6.70)
$$

for $k = 0, 1, 2, \ldots$ $\qquad\qquad\qquad\qquad\qquad\qquad\qquad\qquad\qquad\qquad\square$

6.3   Further Reading

The control theoretic smoothing spline was first proposed in [108]. The book [36] is a nice reference for the smoothing spline. The convex optimization formulation of constrained smoothing spline was considered in [78], and the sparse representation was proposed in [79].

The maximum hands-off control was first proposed in [83] for continuous-time and discrete-time systems. Detailed discussions of maximum hands-off control for continuous-time systems can be found in Part II of this book. The model predictive control formulation was proposed in [86].

Sparsity Methods in Optimal Control

DOI: 10.1561/9781680837254.ch7

Chapter 7

Dynamical Systems and Optimal Control

We have studied the idea and algorithms of sparse optimization in Part I. In part II, we will extend the sparsity methods to dynamical systems. For this, we here review basics of dynamical systems and optimal control, before we consider sparse control in the subsequent chapters.

Key ideas of Chapter 7

- A dynamical system is modeled by a differential equation called the state-space equation.
- We cannot control uncontrollable systems.
- Optimal control is the best control among feasible controls for a controllable system.

7.1 Dynamical System

A *dynamical system* is a system that depends on time. That is, a dynamical system is a *moving* system. Dynamical systems are around us; industrial products such as vehicles, airplanes, motors, electric circuits, etc, as well as movement of planetary, change of weather, ant swarm, cell movement, fluctuations in stock prices and spread of virus. Dynamical systems are important not only in engineering but also in physics, biology, economics, and social science.

7.1.1 State Equation

In this book, we focus on a dynamical system that is described by a linear differential equation:

$$\dot{x}(t) = Ax(t) + bu(t), \quad t \geq 0, \quad x(0) = \xi \in \mathbb{R}^d, \tag{7.1}$$

where $A \in \mathbb{R}^{d \times d}$, $b \in \mathbb{R}^d$, $x(t) \in \mathbb{R}^d$, and $u(t) \in \mathbb{R}$. We call $x(t)$ the *state*, and $u(t)$ the *control*. The state $x(0) = \xi$ at time $t = 0$ is called the *initial state*, and the differential equation in (7.1) is called the *state equation*. The dynamical system in (7.1) is controlled by the control $u(t)$, and called a *controlled object* or a *plant*.

Exercise 7.1. Show that the solution of the differential equation (7.1) is given by

$$x(t) = e^{At}\xi + \int_0^t e^{A(t-\tau)} bu(\tau)d\tau, \quad t \geq 0. \tag{7.2}$$

Example 7.2 (Rocket). *Let us consider the control of a rocket in the outer space where no friction nor gravity acts (see Figure 7.1). The rocket is accelerated by thrust from a rocket engine. Let the mass of the rocket be m [kg]. We assume that the rocket can move on 1-dimensional straight line. Let the position of the rocket at time $t \geq 0$ be $r(t)$ with initial position $r(0) = \xi_1$, and initial velocity $v(0) = \dot{r}(0) = \xi_2$. We denote the thrust force by $F(t)$. From the Newton's second law of motion, we have*[1]

$$m\ddot{r}(t) = F(t), \quad r(0) = \xi_1, \quad \dot{r}(0) = \xi_2. \tag{7.3}$$

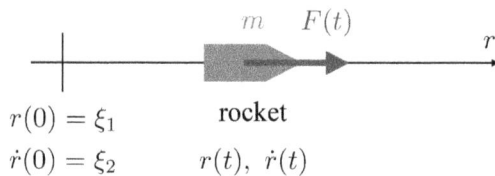

Figure 7.1. Rocket example.

---

1.   Strictly speaking, the thrust of a rocket is obtained by emitting its mass (e.g. fuel) to the opposite direction, and hence the model is not correct. That is, the mass m should be time-varying $m(t)$ that decreases in time. In this example, however, we assume that the mass of the rocket is sufficiently large and the variation can be ignored.

Let us transform this differential equation into the state equation in (7.1). *For this, define the state* $x(t)$ *by*

$$x(t) \triangleq \begin{bmatrix} x_1(t) \\ x_2(t) \end{bmatrix} \triangleq \begin{bmatrix} r(t) \\ \dot{r}(t) \end{bmatrix}. \tag{7.4}$$

Then we have

$$\dot{x}(t) = \begin{bmatrix} \dot{r}(t) \\ \ddot{r}(t) \end{bmatrix} = \begin{bmatrix} x_2(t) \\ m^{-1}F(t) \end{bmatrix} = \begin{bmatrix} 0 & 1 \\ 0 & 0 \end{bmatrix} \begin{bmatrix} x_1(t) \\ x_2(t) \end{bmatrix} + \begin{bmatrix} 0 \\ m^{-1} \end{bmatrix} F(t) \tag{7.5}$$

Defining $u(t) \triangleq F(t)$ *and*

$$A \triangleq \begin{bmatrix} 0 & 1 \\ 0 & 0 \end{bmatrix}, \quad b \triangleq \begin{bmatrix} 0 \\ m^{-1} \end{bmatrix}, \quad \xi \triangleq \begin{bmatrix} \xi_1 \\ \xi_2 \end{bmatrix}, \tag{7.6}$$

we obtain the state equation of the form (7.1).

The system (7.3) or (7.5) is sometimes called the *double integrator*, since the position $r(t)$ is obtained by integrating $F(t)$ twice.

Let us investigate the meaning of the state-space equation (7.1). Assume that the initial state $x(0) = \xi$ at time $t = 0$ is obtained from observation by a sensor attached to the system. The signal $u(t)$ is called a *control* and we design $u(t)$ for $t \geq 0$ to realize a desired trajectory of the state $x(t)$. In the rocket control considered in Example 7.2, we design the thrust force $u(t) = F(t)$ to drive the rocket, for example, within time $T > 0$ from the earth $(x(0) = \xi)$ to the moon $x(T) = 0$ with minimum fuel consumption. This is a problem of control.

If the control $u(t)$ for $t \geq 0$ depends only on the initial state $x(0) = \xi$, then the control is called *feedforward control*. Instead, if the control $u(t)$ for $t \geq 0$ is determined a constant (or an intermittent) observation of the state $x(\tau)$ with $0 \leq \tau \leq t$, then the control is called *feedback control*. Feedforward control uses only one observation $x(0)$ at time $t = 0$. This is, so to speak, driving a bicycle (or a car) with eyes closed, while feedback control uses information from the eyes which is always (or sometimes) open. From this observation, we can easily understand that feedforward control is very fragile against uncertainties and disturbances. The feedback structure solves this fragility and leads to *robustness*. However, we mainly consider feedforward control since it gives clear mathematical structures of the optimal control. For feedback control implementation, one can adopt the receding horizon control, also known as

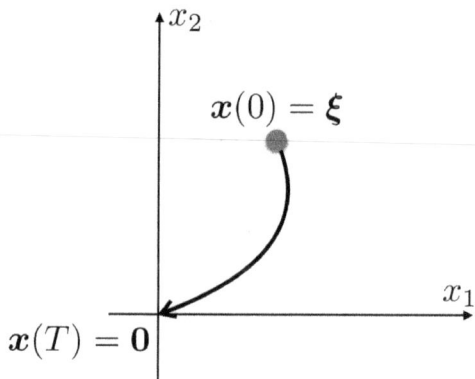

Figure 7.2. State transfer problem: finding a control $u(t)$, $0 \le t \le T$ that drives the state from a given initial state $x(0) = \xi$ to 0.

the model predictive control [70] as discussed in Section 6.2, or self-triggered control [48].

7.1.2   Controllability and Controllable Set

We can consider many types of objectives of controlling the plant (7.1). For example, we set several target points $x_1,\ldots,$ x_s to control the plant so that the state $x(t)$ pass approximately thorough these points at time $t = T_1, \ldots, T_s$, that is $x(T_i) \approx x_i$. This control is called *trajectory generation*, or *trajectory planning*. We can also consider a control problem to keep the state $x(t)$, $t \ge 0$ in a prescribed set $\mathcal{X}$ in the state space, that is, $x(t) \in \mathcal{X}$ for all $t \ge 0$, assuming $x(0) \in \mathcal{X}$. This problem arises for example in keeping a drone hovering in a region.

In this book, we mainly focus on the problem of *state transfer*. This problem is finding a control $u(t)$ that drives the state $x(t)$ from a given initial state ξ to the origin $\mathbf{0}$ in a given time $T > 0$ (see also Figure 7.2).

First, we discuss the existence of the control. For this, we introduce the notion of controllability.

Definition 7.3 (Controllability).   *We call the system* (7.1) *is controllable if for any initial state* $x(0) = \xi \in \mathbb{R}^d$, *there exist a time* $T > 0$ *and a control* $u(t)$, $0 \le t \le T$ *such that the state* $x(t)$ *in* (7.1) *is driven to the origin at time* $t = T$, *that is* $x(T) = \mathbf{0}$.

If the system is not controllable, then there exists an initial state that cannot be achieved to the origin with any $u(t)$ in finite time. The controllability is a fundamental requirement for control systems, and in this book we always assume that the system (7.1) is controllable.

Given a linear system, to check its controllability is an easy task. In fact, we have the following theorem for the controllability:

Theorem 7.4. *The dynamical system* (7.1) *is controllable if and only if any of the following equivalent conditions is satisfied:*

1. *The following matrix called the* controllability matrix

$$M \triangleq \begin{bmatrix} b & Ab & A^2b & \cdots & A^{d-1}b \end{bmatrix} \qquad (7.7)$$

   is non-singular.
2. *The following matrix called the* controllability grammian

$$G(T) \triangleq \int_0^T e^{At} bb^\top e^{A^\top t} dt \qquad (7.8)$$

   is non-singular for any $T > 0$.
3. *For any* $\lambda \in \mathbb{C}$,

$$\text{rank} \begin{bmatrix} A - \lambda I & B \end{bmatrix} = d. \qquad (7.9)$$

4. *For any left eigenvector* $v^\top$ *of* A,

$$v^\top b \neq 0. \qquad (7.10)$$

From this theorem, to check the controllability of the dynamical system (7.1) is just to compute the determinant of the matrix M.

The controllability of the system (7.1) is completely determined by the matrix pair (A, b). From this, we often say the pair (A, b) *is controllable*, which means the system (7.1) is controllable.

Example 7.5. *Let us consider the rocket model* (7.5) *and* (7.6) *in Example 7.2. The controllability matrix is given by*

$$M = \begin{bmatrix} b & Ab \end{bmatrix} = \begin{bmatrix} 0 & 1 \\ 1 & 0 \end{bmatrix}. \qquad (7.11)$$

It is easily shown that this matrix is non-singular. Thus, the system is controllable from Theorem 7.4.

Note that if the dynamical system (7.1) is controllable, then for any initial state $\xi \in \mathbb{R}^d$, any final state $\zeta \in \mathbb{R}^d$, and any time $T > 0$, there exist a control $u(t)$, $0 \leq t \leq T$ that drives the state $x(t)$ from $x(0) = \xi$ to $x(T) = \zeta$.

Exercise 7.6. Prove the above fact.

In general, the shorter the time $T > 0$ is, the larger the magnitude and the shorter the support of $u(t)$ should be. The shape of $u(t)$ may approach to something like the Dirac's delta when T approaches to zero. However, in real systems, the actuator cannot generate arbitrarily large magnitude of control, and there is always a limit on the maximum magnitude (can you make a vehicle that moves at 1000km/h?). Hence, we assume the following limitation on $u(t)$:

$$|u(t)| \leq 1, \quad \forall t \in [0, T]. \tag{7.12}$$

We call a control that satisfies this constraint an *admissible control*. In (7.12), we assume the maximum magnitude is normalized to one, but if the maximum magnitude is $U_{max} > 0$ and the limitation is represented by

$$|u(t)| \leq U_{max}, \quad \forall t \in [0, T], \tag{7.13}$$

then, we can redefine the vector b in the plant (7.1) as

$$b' \triangleq \frac{b}{U_{max}}, \tag{7.14}$$

then the limitation is reduced to (7.13).

Under the constraint (7.12), there may be an initial state ξ that cannot be steered to the origin by any admissible control $u(t)$ that satisfies (7.12) within time $T > 0$ even if the system is controllable. To discuss this, we introduce the notion of the T-controllable set:

Definition 7.7 (T-Controllable Set). *Fix $T > 0$. The set of initial states that can be steered to the origin by some admissible control $u(t)$, $0 \leq t \leq T$ is called the T-controllable set. We denote this set by $\mathcal{R}(T)$.*

Exercise 7.8. Prove that $\mathcal{R}(T)$ can be represented by

$$\mathcal{R}(T) = \left\{ - \int_0^T e^{-At} bu(t)dt : |u(t)| \leq 1, \ \forall t \in [0, T] \right\}. \tag{7.15}$$

For the T-controllable set, we have the following theorem:

Theorem 7.9. *For any $T > 0$, the T-controllable set $\mathcal{R}(T)$ is a bounded, closed, and convex set. Also, if $T_1 < T_2$ then $\mathcal{R}(T_1) \subset \mathcal{R}(T_2)$.*

Exercise 7.10. Prove Theorem 7.9.

Figure 7.3 shows an illustration of a T-controllable set $\mathcal{R}(T)$ in $\mathbb{R}^2$. If an initial state $x(0) = \xi$ is in the T-controllable set $\mathcal{R}(T)$, then there exists an admissible control $u(t)$, $0 \leq t \leq T$ that steers the state to $x(T) = 0$ in time T. If an initial state is outside the set $\mathcal{R}(T)$, then such control does not exist. We show an easy example to illustrate this property.

Example 7.11. *Let us consider control of a ball on an inclined plane shown in Figure 7.4. Let $x(t)$ denote the position of the ball on the x axis parallel to the slope.*

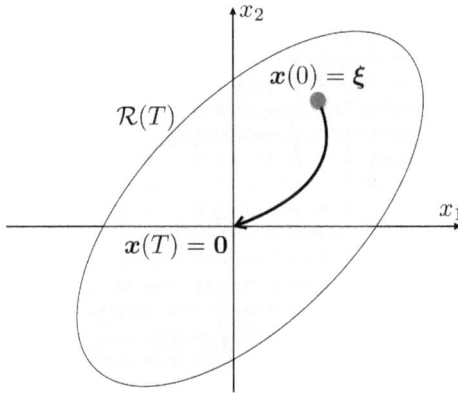

Figure 7.3. T-Controllable set $\mathcal{R}(T)$ in $\mathbb{R}^2$.

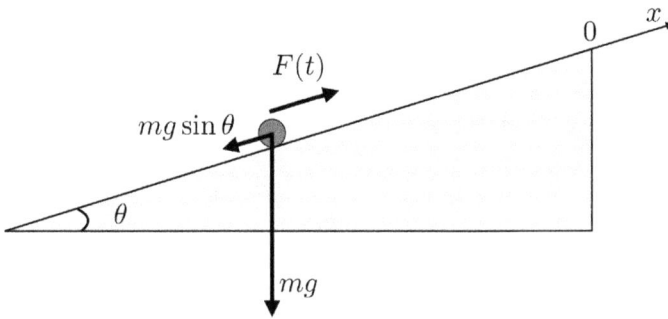

Figure 7.4. A ball on an inclined plane: m is the mass, g is the acceleration of gravity, θ is the angle of the slope, and $F(t)$ is the force (i.e., control) applied to the ball.

The origin is set at the top of the slope. The control objective is to move the ball from the initial position $x(0) = \xi$ to the origin within time $T > 0$, that is, $x(T) = 0$.
    The differential equation of $x(t)$ is given from Newton's second law of motion:

$$m\ddot{x}(t) = F(t) - mg \sin\theta. \tag{7.16}$$

Now, we assume

$$x(0) = -\xi, \quad \dot{x}(0) = 0, \tag{7.17}$$

where $\xi > 0$. From (7.16), we have

$$\dot{x}(t) = \frac{1}{m} \int_0^t F(\tau)d\tau - gt \sin\theta + \dot{x}(0), \tag{7.18}$$

and

$$x(t) = \frac{1}{m} \int_0^t \int_0^s F(\tau)d\tau ds - \frac{1}{2}gt^2 \sin\theta + \dot{x}(0)t + x(0). \tag{7.19}$$

Then, if there exists an admissible control $\{F(t) : 0 \leq t \leq T\}$ such that $x(T) = 0$, then we have

$$0 = x(T) = \frac{1}{m} \int_0^T \int_0^s F(\tau) d\tau ds - \frac{1}{2} g T^2 \sin\theta - \xi, \qquad (7.20)$$

where we used the initial conditions in (7.17). From the above equation, we have

$$\left| \frac{1}{2} g T^2 \sin\theta + \xi \right| \leq \int_0^T \int_0^s |F(\tau)| d\tau ds$$

$$\leq \int_0^T \int_0^s \|F\|_\infty d\tau ds \qquad (7.21)$$

$$= \frac{T^2}{2m} \|F\|_\infty,$$

where $\|F\|_\infty$ is the L^∞ norm of function F defined by

$$\|F\|_\infty \triangleq \sup_{t \in [0,T]} |F(t)|. \qquad (7.22)$$

Since the variables m, g, T, and ξ are all positive, and $\sin\theta$ is also positive, we have

$$\|F\|_\infty \geq mg \sin\theta + \frac{2m\xi}{T^2}. \qquad (7.23)$$

On the other hand, since F is an admissible control, it should satisfy

$$\|F\|_\infty \leq 1. \qquad (7.24)$$

From (7.23) and (7.24), we have

$$1 \geq mg \sin\theta + \frac{2m\xi}{T^2}. \qquad (7.25)$$

It follows that $mg \sin\theta < 1$ and

$$T \geq \sqrt{\frac{2m\xi}{1 - mg \sin\theta}} \triangleq T^*. \qquad (7.26)$$

From this, if the final time T is small such that $T < T^*$, then there is no admissible control with time T that achieves $x(T) = 0$. Conversely, let $T = T^*$ and take

$F(t) \equiv 1, 0 \leq t \leq T^*$. *Then from (7.19), we can easily compute that*

$$x(T^*) = 0. \tag{7.27}$$

Hence, $F(t) \equiv 1$ is a solution with $T = T^*$. Also, if $T > T^*$, then if we choose $F(t)$ as

$$F(t) = \begin{cases} 1, & \text{if } 0 \leq t \leq T^*, \\ mg \sin\theta, & \text{if } T^* < t \leq T, \end{cases} \tag{7.28}$$

then you can easily show that this is a solution with time T.

From this example, the time T^* is the threshold that determines the T-controllability with a fixed initial state. The time T^* is called the *minimum time*, which is in general defined by

$$T^*(\xi) \triangleq \inf\{T \geq 0 : \xi \in \mathcal{R}(T)\}. \tag{7.29}$$

To consider the minimum time, we define the *controllable set* by the union of all $\mathcal{R}(T)$ with $T > 0$, that is,

$$\mathcal{R} \triangleq \cup_{T>0} \mathcal{R}(T). \tag{7.30}$$

Even if the system is controllable, the controllable set $\mathcal{R}$ may not be $\mathbb{R}^d$. That is, $\mathcal{R}$ may be a strict subset of $\mathbb{R}^d$. Then, if $\xi \notin \mathcal{R}$, then there exists no admissible control on any finite time interval $[0, T]$ that achieves $x(T) = \mathbf{0}$. For this case, we write $T^*(\xi) = \infty$. Conversely, if $\xi \in \mathcal{R}$ then the set $\{T \geq 0 : \xi \in \mathcal{R}(T)\}$ is non-empty, and the minimum time (7.29) is finite, that is, $T^*(\xi) < \infty$.

Assume that $\xi \in \mathcal{R}$. Then $T^*(\xi) < \infty$. Let us consider T_1 and T_2 such that

$$T_1 < T^*(\xi) < T_2. \tag{7.31}$$

Then we have

$$\mathcal{R}(T_1) \subset \mathcal{R}(T^*(\xi)) \subset \mathcal{R}(T_2) \subset \mathcal{R}. \tag{7.32}$$

This inclusion relation is shown in Figure 7.5. From this figure, we have $\xi \in \mathcal{R}(T_2)$ and $\xi \notin \mathcal{R}(T_1)$. This means that if the final time is greater than $T^*(\xi)$, then there exists a feasible control, while if it is less than $T^*(\xi)$, then there is no control. The minimum time $T^*(\xi)$ is the threshold for the controllability, that is, ξ is on the boundary of the $T^*(\xi)$-controllability set $\mathcal{R}(T^*(\xi))$. We will discuss the minimum time further in the next subsection.

For a stable system, the minimum time always exists for any initial state. In fact, we have the following theorem [49, Theorem 17.6]:

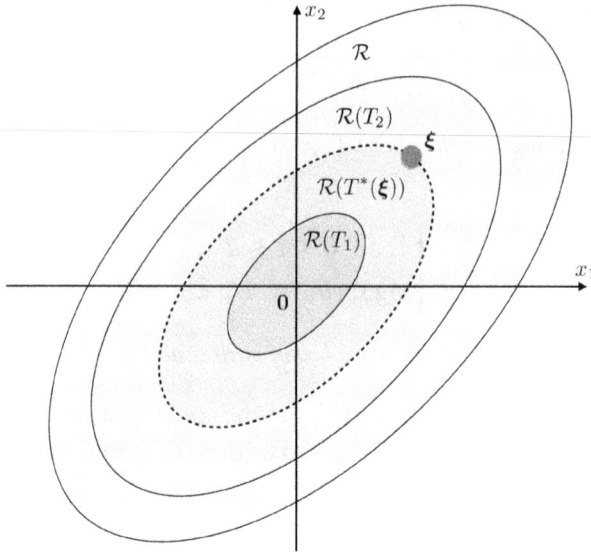

Figure 7.5. Controllable sets $\mathcal{R}(T_1) \subset \mathcal{R}(T^*(\xi)) \subset \mathcal{R}(T_2) \subset \mathcal{R}$, where $T_1 < T^*(\xi) < T_2$. The minimum time $T^*(\xi)$ is the threshold for the feasibility from the initial state, ξ.

Theorem 7.12. *Assume that (A, b) is controllable. Assume also that A is stable, that is,*

$$\lambda(A) \subset \mathbb{C}_- \triangleq \{z \in \mathbb{C} : \operatorname{Re} z \leq 0\}, \tag{7.33}$$

where $\lambda(A)$ is the set of eigenvalues of A. Then the controllable set $\mathcal{R}$ is $\mathbb{R}^d$, and the minimum time $T^*(\xi)$ is finite for any $\xi \in \mathbb{R}^d$.

7.1.3   Feasible Control and Minimum-time Control

Fix $T > 0$ and assume $x(0) = \xi \in \mathcal{R}(T)$. Then by the definition of controllable set in Definition 7.7, there exists an admissible control $u(t)$ that steers the state from $x(0)$ to $x(T) = 0$. We call such a control a *feasible control*. Let denote by $\mathcal{U}(T, \xi)$ the set of feasible controls with initial state ξ and final time T. This set can be represented by

$$\mathcal{U}(T, \xi) = \left\{ u \in L^\infty(0, T) : \xi = -\int_0^T e^{-At} bu(t) dt, \ |u(t)| \leq 1, \forall t \in [0, T] \right\}. \tag{7.34}$$

Exercise 7.13. Prove that the set of feasible controls is represented by (7.34).

It is easily shown that $\xi \in \mathcal{R}(T)$ if and only if there exists an admissible control u such that $u \in \mathcal{U}(T, \xi)$. Hence the minimum time $T^*(\xi)$ in (7.29) is formulated by

$$T^*(\xi) = \inf\{T \geq 0 : \exists u, \ u \in \mathcal{U}(T, \xi)\}. \tag{7.35}$$

From the discussion in the previous subsection, $T^*(\xi)$ is finite if and only if $\xi \in \mathcal{R}$. From this, if $\xi \in \mathcal{R}$, then there exists a finial time $T \geq 0$ and an admissible control u such that $u \in \mathcal{U}(T, \xi)$, and hence the set of the right hand side of (7.35) is non-empty.

Now we find the control that achieves this minimum time. That is, we consider the following optimization problem:

$$\underset{u}{\text{minimize}} \ T \ \text{subject to} \ u \in \mathcal{U}(T, \xi). \tag{7.36}$$

The solution is called the *minimum time control* or the *time-optimal control*. The minimum control exists if $\xi \in \mathcal{R}$ or equivalently $T^*(\xi) < \infty$. In fact, we have the following lemma:

Theorem 7.14. *Assume* $T^*(\xi) < \infty$. *Then there exists a minimum-time control* $u^* \in \mathcal{U}(T^*(\xi), \xi)$. *Moreover, for any* $T > T^*(\xi)$, $\mathcal{U}(T, \xi)$ *is non-empty.*

Exercise 7.15. Prove Theorem 7.14.

7.1.4 Optimal Control and Pontryagin Minimum Principle

From Theorem 7.14, if $\xi \in \mathcal{R}$ and $T > T^*(\xi)$, then the set of feasible controls $\mathcal{U}(T, \xi)$ is non-empty, and in general $\mathcal{U}(T, \xi)$ has infinitely many elements. *Optimal control* is the optimal one with a given cost function among all feasible controls in $\mathcal{U}(T, \xi)$.

The following is the formulation of the optimal control that is mainly considered in this book:

Optimal Control Problem (OPT)

For the plant modeled by

$$\dot{x}(t) = Ax(t) + bu(t), \quad t \geq 0, \quad x(0) = \xi \in \mathbb{R}^d,$$

find an admissible control u (i.e. $\|u\|_\infty \leq 1$) that achieves

$$x(T) = 0,$$

and minimizes the following cost function:

$$J(u) = \int_0^T L\big(u(t)\big)dt.$$

We here assume that the function $L(u)$, called the *stage cost function*, is continuous in u. We call the solution the *optimal control*. Note that the optimal control problem can also be written by using the set of feasible controls $\mathcal{U}(T, \xi)$ as

$$\underset{u}{\text{minimize}} \ J(u) \ \text{subject to} \ u \in \mathcal{U}(T, \xi). \tag{7.37}$$

The minimum-time control (7.36) is the optimal control with $L(u) = 1$.

Let us assume that there exists an optimal control for the problem (OPT). We here introduce *Pontryagin's minimum principle* that gives necessary conditions for the optimal control.

First, define the following function called *Hamiltonian*:

$$H^\eta(x, p, u) \triangleq p^\top (Ax + bu) + \eta L(u), \tag{7.38}$$

where $\eta \in \{0, 1\}$ is called the *abnormal multiplier*. The following theorem is Pontryagin's minimum principle.

Theorem 7.16 (Pontryagin's Minimum Principle (PMP)). *Assume that an optimal control u^* of the optimal control problem (OPT) exists. Let us denote by $\{x^*(t) : 0 \le t \le T\}$ the optimal state with the optimal control $\{u^*(t) : 0 \le t \le T\}$, that is,[2]*

$$x^*(t) \triangleq e^{At}\xi + \int_0^t e^{A(t-\tau)}bu^*(\tau)d\tau, \quad \forall t \in [0, T]. \tag{7.39}$$

Then there exist $\eta \in \{0, 1\}$ and the optimal costate $\{p^*(t) : 0 \le t \le T\}$ that satisfy the following conditions.

(non-triviality condition) *The abnormal multiplier η and the optimal costate p^* satisfy the non-triviality condition:*

$$|\eta| + \|p^*\|_\infty > 0. \tag{7.40}$$

(canonical equation) *The following canonical equations hold*

$$\dot{x}^*(t) = Ax^*(t) + bu^*(t),$$
$$\dot{p}^*(t) = -A^\top p^*(t), \quad \forall t \in [0, T]. \tag{7.41}$$

The differential equation for $p^*(t)$ is called the adjoint equation.

---

2.   See the solution formula (7.2) for the differential equation $\dot{x} = Ax + bu$.

(minimum condition) *The optimal control $u^*(t)$ minimizes Hamiltonian at each time $t \in [0, T]$. That is,*

$$u^*(t) = \arg\min_{|u| \leq 1} H^{\eta}(x^*(t), p^*(t), u), \quad \forall t \in [0, T]. \tag{7.42}$$

(consistency) *Hamiltonian satisfies*

$$H^{\eta}(x^*(t), p^*(t), u^*(t)) = c, \quad \forall t \in [0, T], \tag{7.43}$$

where c is a constant independent of t. If T is not fixed (as in the minimum-time control), then

$$H^{\eta}(x^*(t), p^*(t), u^*(t)) = 0, \quad \forall t \in [0, T]. \tag{7.44}$$

Note that the canonical equation in (7.41) can be rewritten in terms of Hamiltonian H^{η} as

$$\dot{x}^*(t) = \frac{\partial H^{\eta}}{\partial p}(x^*(t), p^*(t), u^*(t))$$

$$\dot{p}^*(t) = -\frac{\partial H^{\eta}}{\partial x}(x^*(t), p^*(t), u^*(t)), \quad \forall t \in [0, T]. \tag{7.45}$$

These equations are also called *Hamilton's canonical equations*.

Pontryagin's minimum principle is a powerful tool to analyze the optimal control (if it exists). For simple problems, we can obtain a closed form of the control that satisfies the necessary conditions. We call this an *extremal control*. We should note that an extremal control is not necessarily the optimal control. However, in some cases, we can determine the optimal control from the minimum principle. One example is shown in Section 7.3, the minimum-time control for the rocket in Example 7.2. Before the example, we will formulate the minimum-time control for general linear systems.

7.2   Minimum-time Control

Let us consider the following linear system:

$$\dot{x}(t) = Ax(t) + bu(t), \quad t \geq 0, \quad x(0) = \xi \in \mathbb{R}^d. \tag{7.46}$$

For this system, we consider the minimum-time control, which is given by the optimal control (OPT) with the stage cost $L(u) = 1$. Then the Hamiltonian is given by

$$H^{\eta}(x, p, u) = p^{\top}(Ax + bu) + \eta. \tag{7.47}$$

Let us assume that the minimum-time control exists. Then from Pontryagin's minimum principle, the optimal control $u^*(t)$ should satisfy

$$u^*(t) = \underset{u \in [-1,1]}{\arg\min}\, H^\eta(x^*(t), p^*(t), u), \quad \forall t \in [0, T^*(\xi)], \tag{7.48}$$

where $x^*(t)$ and $p^*(t)$ are respectively the optimal state and costate by the optimal control $u^*(t)$, and $T^*(\xi)$ is the minimum time. From this, we have

$$u^*(t) = \underset{u \in [-1,1]}{\arg\min}\, p^*(t)^\top bu = -\mathrm{sgn}(p^*(t)^\top b), \tag{7.49}$$

where $\mathrm{sgn}(\cdot)$ is the *sign function* defined by

$$\mathrm{sgn}(a) = \begin{cases} 1, & a > 0 \\ -1, & a < 0 \end{cases} \tag{7.50}$$
$$\mathrm{sgn}(a) \in [-1, 1], \quad a = 0.$$

If the function $p^*(t)^\top b$ is not zero for almost all $t \in [0, T^*(\xi)]$, then the control $u^*(t)$ takes values of only ± 1 for almost all t. Such a control is called a *bang-bang control*.

Lemma 7.17. *If (A, b) is controllable, then the function $p^*(t)^\top b$ is not zero for almost all $t \in [0, T^*(\xi)]$.*

Exercise 7.18. Prove Lemma 7.17.

For the minimum-time control problem, we have the following existence and uniqueness theorems.

Theorem 7.19 (Existence). *If the initial state ξ is in the controllable set $\mathcal{R}$ defined in (7.30), then a minimum-time control exists.*

Theorem 7.20 (Uniqueness). *Assume that (A, b) is controllable. Then the minimum-time control is (if it exists) unique.*

Exercise 7.21. Prove Theorems 7.19 and 7.20.

The following corollary is easily proved from Theorems 7.12, 7.19, and 7.20.

Corollary 7.22. *Assume that (A, b) is controllable and A is stable. Then for any $\xi \in \mathbb{R}^d$, the minimum-time control $u^* \in \mathcal{U}(\xi)$ uniquely exists.*

7.3 Minimum-time Control of Rocket

Here we derive the minimum-time control of the rocket in Example 7.2 by using Pontryagin's minimum principle.

Now, from Example 7.5, the pair (A, b) is controllable. It is also easily seen that A is stable since A has a multiple eigenvalue of 0. Therefore, from Theorem 7.22, there uniquely exists the minimum-time control u^*.

Let us define the optimal state and costate by

$$\boldsymbol{x}^*(t) = \begin{bmatrix} x_1^*(t) \\ x_2^*(t) \end{bmatrix}, \quad \boldsymbol{p}^*(t) = \begin{bmatrix} p_1^*(t) \\ p_2^*(t) \end{bmatrix}. \tag{7.51}$$

For simplicity, we assume the mass of the rocket $m = 1$.

Then the Hamiltonian $H^\eta(\boldsymbol{x}, \boldsymbol{p}, u)$ in (7.47) is given by

$$H^\eta(\boldsymbol{x}, \boldsymbol{p}, u) = \boldsymbol{p}^\top(A\boldsymbol{x} + bu) + \eta = p_1 x_2 + p_2 u + \eta. \tag{7.52}$$

The canonical equation (7.41) for the costate $\boldsymbol{p}^*(t)$ is given by

$$\begin{aligned} \dot{p}_1^*(t) &= 0, \\ \dot{p}_2^*(t) &= -p_1^*(t). \end{aligned} \tag{7.53}$$

Let $p_1^*(0) = \pi_1$ and $p_2^*(0) = \pi_2$. Then the solution to the differential equation (7.53) is given by

$$\begin{aligned} p_1^*(t) &= \pi_1, \\ p_2^*(t) &= \pi_2 - \pi_1 t. \end{aligned} \tag{7.54}$$

Since T is not fixed, from the condition (7.44), we have $H^\eta(\boldsymbol{x}^*(t), \boldsymbol{p}^*(t), u^*) = 0$. That is,

$$p_1^*(t)x_2^*(t) + p_2^*(t)u^* + \eta = 0. \tag{7.55}$$

If $\pi_1 = \pi_2 = 0$, then $p_1^*(t) = p_2^*(t) = 0$ from (7.54), and hence $\eta = 0$ from (7.55). But this contradicts the non-triviality condition (7.40). Therefore, $\pi_1 \neq 0$ or $\pi_2 \neq 0$, that is $\boldsymbol{p}^*(0) \neq \boldsymbol{0}$.

Next, from (7.49), we have

$$u^*(t) = -\mathrm{sgn}\left(\boldsymbol{p}^*(t)^\top b\right) = -\mathrm{sgn}\left(p_2^*(t)\right). \tag{7.56}$$

From (7.54), $p_2^*(t)$ is a linear function $p_2^*(t) = \pi_2 - \pi_1 t$. Then we need to check the following cases:

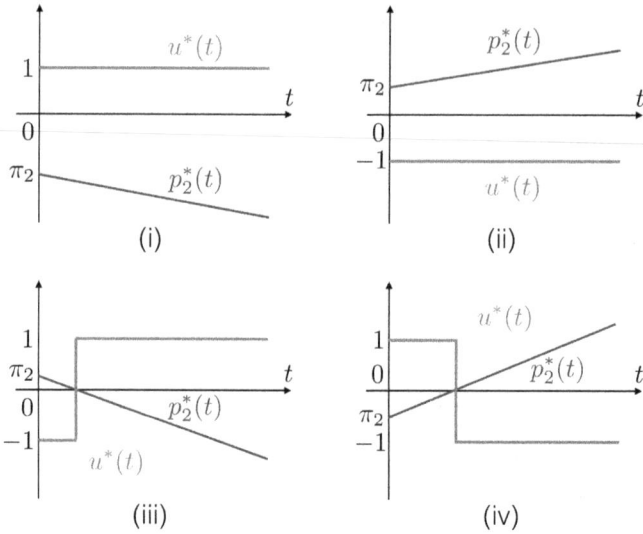

Figure 7.6. Costate $p_2^*(t) = \pi_2 - \pi_1 t$ and corresponding extremum control $u^*(t)$ from (7.56).

(i) $\pi_1 \leq 0, \pi_2 \leq 0$ with $(\pi_1, \pi_2) \neq (0, 0)$

(ii) $\pi_1 \geq 0, \pi_2 \geq 0$ with $(\pi_1, \pi_2) \neq (0, 0)$

(iii) $\pi_1 < 0, \pi_2 > 0$

(iv) $\pi_1 > 0, \pi_2 < 0$

Extremum controls given in (7.56) for the 4 cases are shown in Figure 7.6. From this figure, it is easily shown that each extremum control takes its values of ± 1 for almost all t, that is bang-bang. We note that the number of switching is at most one from Figure 7.6.

Next, we compute the trajectory $x(t)$ when $u(t)$ is constant (i.e. ± 1). From (7.5), we have

$$\dot{x}_1(t) = x_2(t), \quad x_1(0) = \xi_1,$$
$$\dot{x}_2(t) = u(t), \quad x_2(0) = \xi_2. \tag{7.57}$$

If $u(t) = \pm 1$, then

$$x_1(t) = \pm \frac{1}{2}t^2 + \xi_2 t + \xi_1,$$
$$x_2(t) = \pm t + \xi_2. \tag{7.58}$$

Eliminating the time variable t gives

$$x_1(t) = \pm \frac{1}{2}x_2(t)^2 + \xi_1 \mp \frac{1}{2}\xi_2^2. \tag{7.59}$$

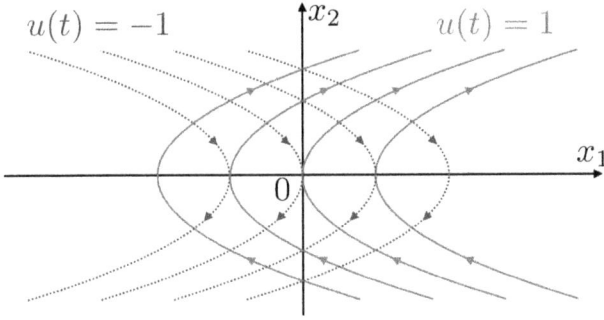

Figure 7.7. Flow of state $(x_1(t), x_2(t))$ by constant control $u(t) = 1$ (solid curve) and $u(t) = -1$ (dashed curve).

That is, when the control $u(t)$ is constant ± 1, then the state $(x_1(t), x_2(t))$ moves on the following parabolic curve:

$$x_1 = \frac{1}{2}x_2^2 + \xi_1 - \frac{1}{2}\xi_2^2, \qquad \text{if } u(t) = 1, \tag{7.60}$$

$$x_1 = -\frac{1}{2}x_2^2 + \xi_1 + \frac{1}{2}\xi_2^2, \qquad \text{if } u(t) = -1. \tag{7.61}$$

Figure 7.7 shows the flow of the state $(x_1(t), x_2(t))$ by some of these parabolic curves with directions of the state to move. Note that the parabolic curves defined in (7.60) and (7.61) go through the point (ξ_1, ξ_2).

To achieve the terminal state $x(T) = 0$, the final trajectory must on the parabolic curve that go through the origin:

$$x_1 = \frac{1}{2}x_2^2, \qquad \text{if } u(t) = 1,$$
$$x_1 = -\frac{1}{2}x_2^2, \qquad \text{if } u(t) = -1. \tag{7.62}$$

From this, if there is no switching, that is, in the cases of (see Figure 7.7)

(i) $u^*(t) \equiv 1$
(ii) $u^*(t) \equiv -1$

then, the initial state (ξ_1, ξ_2) should be on the parabolic curve

$$\gamma_+ \triangleq \left\{ (x_1, x_2) \in \mathbb{R}^2 : x_1 = x_2^2/2, \ x_2 \le 0 \right\} \tag{7.63}$$

or

$$\gamma_- \triangleq \left\{ (x_1, x_2) \in \mathbb{R}^2 : x_1 = -x_2^2/2, \ x_2 \le 0 \right\}. \tag{7.64}$$

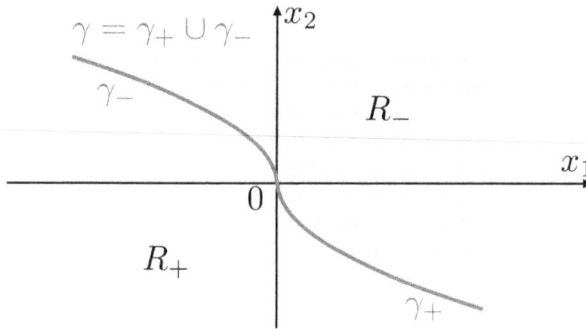

Figure 7.8. Switching curve $\gamma = \gamma_+ \cup \gamma_-$ and regions R_+ and R_-.

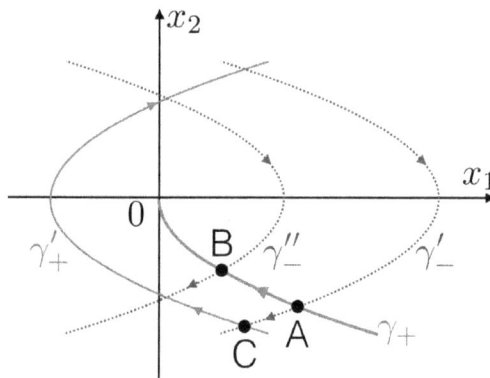

Figure 7.9. 4 cases of state trajectories from initial state (ξ_1, ξ_2) on the curve γ_+.

Figure 7.8 shows the two curves γ_+ and γ_-. In fact, we can easily show that

- If the initial state (ξ_1, ξ_2) is on the curve γ_+, then $u^*(t) \equiv 1$ is the unique extremum control.
- If the initial state (ξ_1, ξ_2) is on the curve γ_-, then $u^*(t) \equiv -1$ is the unique extremum control.

The proof is shown below.

Assume $(\xi_1, \xi_2) \in \gamma_+$. As mentioned above, there are 4 extremum controls with (i)–(iv). Now, $u^*(t) \equiv 1$ is for the case (i). The point A in Figure 7.9 is the initial point, and the state can reach the origin by $u^*(t) \equiv 1$ through the curve γ_+. However, for the other cases (ii), (iii), and (iv), the state never reaches the origin from the initial point A. For the case (ii), by the control $u^*(t) \equiv -1$, the state starts at A on the curve γ'_- to the direction to C, and never reaches the origin. For the case (iii), the state moves on the curve γ'_- from A to C by the control $u^*(t) = -1$, which is switched to $u^*(t) = +1$ at C. Then the state moves on the curve γ'_+,

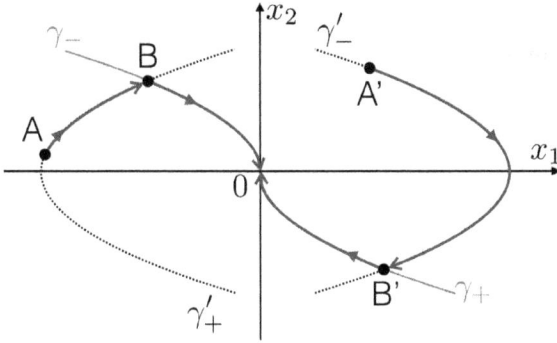

Figure 7.10. State trajectories from initial points A and A'.

which never reaches the origin. Finally, for the case (iv), the state starts from A to B on the curve γ'_+ by the control $u^*(t) = +1$, which is switched to $u^*(t) = -1$ at B. The state then moves from B on the curve γ''_- to the indicated direction and never reaches the origin. In summary, (i) $u^*(t) \equiv 1$ is the unique extremum control, and hence if the minimum-time control exists, this is actually the optimal control. The same discussion can be applied for the initial state (ξ_1, ξ_2) on the curve γ_-, and the unique extremum control is $u^*(t) \equiv -1$.

Next, let us consider the initial state (ξ_1, ξ_2) is outside the curve

$$\gamma \triangleq \gamma_+ \cup \gamma_- = \{(x_1, x_2) \in \mathbb{R}^2 : x_1 = -x_2|x_2|/2\}. \tag{7.65}$$

Let us define two regions R_+ and R_- divided by the curve γ:

$$\begin{aligned} R_+ &\triangleq \{(x_1, x_2) \in \mathbb{R}^2 : x_1 < -x_2|x_2|/2\}, \\ R_- &\triangleq \{(x_1, x_2) \in \mathbb{R}^2 : x_1 > -x_2|x_2|/2\}. \end{aligned} \tag{7.66}$$

Figure 7.8 shows these regions. We call the curve γ the *switching curve*.

Now assume the initial state (ξ_1, ξ_2) is at A in R_+ as in Figure 7.10. From the point A, the curve γ'_+ defined in (7.60) is plotted. By a constant control $u(t) \equiv 1$, the state moves on the curve γ'_+ to the indicated direction from A. At some time, the state touches the switching curve γ_- at B, and the control is switched to $u(t) = -1$. From B, the state goes on the switching curve to the origin. The control switched from $+1$ to -1 is the control in (iii) in Figure 7.6. We can easily show that this is the unique extremum control from any initial point in R_+, in a similar way to the case where the initial state is on the curve $\gamma = \gamma_+ \cup \gamma_-$.

Then, let us consider the initial state $(\xi_1, \xi_2) \in R_-$ at A' in Figure 7.10. First, by the constant control $u(t) = -1$, the state moves on the curve γ'_- from A' to B'. Then the control is switched from -1 to $+1$, and the state is steered to the origin along the curve γ_+. This control is for the case (iv) in Figure 7.6, and actually this is the unique extremum solution.

In summary, the extremum control $u^*(t)$ of the minimum-time control problem is given by

$$u^*(t) = \begin{cases} 1, & \text{if } x(t) \in \gamma_+ \cup R_+ \setminus \{0\}, \\ -1, & \text{if } x(t) \in \gamma_- \cup R_- \setminus \{0\}, \\ 0, & \text{if } x(t) = 0. \end{cases} \tag{7.67}$$

The control $u^*(t)$ depends on the state $x(t)$, and hence the control is a *feedback control*, which changes its value (± 1 or 0) based on the observation of the state $x(t)$.

Exercise 7.23. Compute the minimum time $T^*(\xi)$ from $\xi = (\xi_1, \xi_2)$ to the origin, and the switching time when $\xi \in R_+$ and $\xi \in R_-$.

7.4 Further Reading

For the basics of control theory with state-space formulations, I recommend a nice book by Ogata [89]. For the controllable set, see [103]. The proof of Pontryagin's minimum principle is found in [66]. Pontryagin's minimum principle is also referred to as Pontryagin's maximum principle, which is mathematically equivalent to the minimum principle. The book by Clarke [24] is one of the most reliable books on the Pontryagin's maximum principle. For the minimum-time control, see the classical books of [2, 49, 96].

DOI: 10.1561/9781680837254.ch8

Maximum Hands-off Control

In this chapter, we introduce a new optimal control problem called maximum hands-off control, which is the sparsest control among all feasible controls.

Key ideas of Chapter 8

- Maximum hands-off control is described as L^0-optimal control.
- Under the assumption of non-singularity, L^0-optimal control is equivalent to L^1-optimal control.
- Maximum hands-off control is a ternary signal that takes values of ± 1 and 0. Such a ternary control is called a bang-off-bang control.

8.1 L^0 Norm and Sparsity

Here we introduce mathematical preliminaries for maximum hands-off control. Define the *support* of a function $u(t)$ on a finite interval $[0, T]$ by

$$\text{supp}(u) \triangleq \{t \in [0, T] : u(t) \neq 0\}. \tag{8.1}$$

By using the support, we define the L^0 *norm* by the length of the support of function u, that is,

$$\|u\|_0 \triangleq \mu\big(\text{supp}(u)\big), \tag{8.2}$$

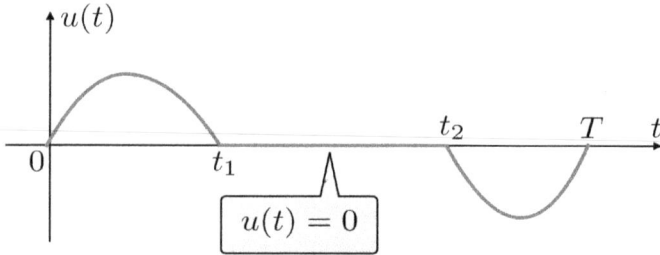

Figure 8.1. The L^0 norm of the function $u(t)$ is $t_1 + (T - t_2)$.

where $\mu(S)$ is the Lebesgue measure of a subset $S \subset [0, T]$. From this definition, the L^0 norm of a continuous-time signal is the total length of time duration on which the signal takes nonzero values.

Example 8.1. *Let us consider a function $u(t)$ in Figure 8.1. The function $u(t)$ is zero over the interval $[t_1, t_2]$, and the support set of u is*

$$\text{supp}(u) = (0, t_1) \cup (t_2, T) \subset [0, T]. \tag{8.3}$$

From this, the L^0 norm of u is

$$\|u\|_0 = \mu(\text{supp}(u)) = t_1 + (T - t_2) = T - (t_2 - t_1). \tag{8.4}$$

In the above example, the value $t_2 - t_1$ is the length of the interval $[t_1, t_2]$ on which $u(t) = 0$. If $\|u\|_0$ is much smaller than the total length T (i.e., $\|u\|_0 \ll T$), then the signal is said to be *sparse*. This notion is an analogy of the sparsity of vectors studied in Part I of this book.

Define

$$|u|^0 \triangleq \begin{cases} 0, & \text{if } u = 0, \\ 1, & \text{if } u \neq 0, \end{cases} \tag{8.5}$$

then the L^0 norm in (8.2) can be written as

$$\|u\|_0 = \int_0^T |u(t)|^0 dt. \tag{8.6}$$

Note that the L^0 norm does not have the absolute homogeneous property (see Definition 2.7, p. 21). In fact, if we take a non-zero scalar α such that $|\alpha| \neq 1$, then

$$\|\alpha u\|_0 = \|u\|_0 \neq |\alpha| \|u\|_0. \tag{8.7}$$

Note also that a sparse signal $u(t)$ on $[0, T]$ has a time duration whose length is positive, on which the control $u(t)$ is exactly zero. This means that the function $u(t)$ is not a real analytic function.[1] For example, a polynomial function

$$p(t) = t^n + a_{n-1}t^{n-1} + \cdots + a_1 t + a_0, \qquad (8.8)$$

a trigonometric function $\sin(\omega t)$ $(\omega \neq 0)$, an exponential function $e^{\lambda t}$, and their sum or product are never sparse.

8.2   Practical Benefits of Sparsity in Control

Let us consider the sparse control signal $u(t), t \in [0, T]$ shown in Figure 8.1. This control signal is exactly zero on the time interval $[t_1, t_2]$. In an electromechanical system, the control signal is transformed into a mechanical motion by an actuator. An electric motor is an example, which transforms the control signal given as an electric current into torque. Usually, an amplifier is attached between a controller and an actuator to supply energy to the actuator enough to generate a mechanical motion. Hence for actuation, we need not only a control signal but also enough energy.

By using a sparse signal as in Figure 8.1, we can stop energy supply to the actuator over the time interval $[t_1, t_2]$. That is, we can save consumption of electric power or fuel over this interval. We call such a control a *hands-off control*, which is also known as *gliding* or *coasting*. This control strategy is actually used in practical control systems. A stop-start system [34, 64] in automobiles is an example of hands-off control. It automatically shuts down the engine to avoid it idling for a long duration of time. By this, we can reduce CO or CO2 emissions as well as fuel consumption. Also in hybrid vehicles [17, 87, 105], the internal combustion engine is stopped when the vehicle is at a stop or the speed is lower than a preset threshold, and the electric motor is alternatively used. Other examples are found in railway vehicles [18, 63] and free-flying robots [113]. Hands-off control is also desirable for networked and embedded systems since the communication can be stopped during a period of zero-valued control. This property is advantageous in particular for wireless communications [60]. By these properties hands-off control is also known as *green control*.

---

1.   A function $u(t)$ is said to be *real analytic* if it is an infinitely differentiable function such that the Taylor series at any point $t_0 \in (0, T)$ converges to $u(t)$ for t in a neighborhood of t_0 pointwise. See [98, Chapter 8] for details.

8.3   Problem Formulation of Maximum Hands-off Control

Let us consider the following linear time-invariant system:

$$\dot{x}(t) = Ax(t) + bu(t), \quad t \geq 0, \quad x(0) = \xi \in \mathbb{R}^d, \tag{8.9}$$

For this system, we consider the optimal control problem (OPT) (p. 141) with the stage cost function

$$L(u) = |u|^0. \tag{8.10}$$

Namely, we seek the optimal control that minimizes the L^0 cost function

$$J_0(u) \triangleq \|u\|_0 = \int_0^T |u(t)|^0 dt, \tag{8.11}$$

among all feasible controls. We call this problem an L^0-*optimal control problem* or a *maximum hands-off control problem*.

L^0-optimal control problem (L^0 OPT)

For the linear time-invariant system

$$\dot{x}(t) = Ax(t) + bu(t), \quad t \geq 0, \quad x(0) = \xi \in \mathbb{R}^d,$$

find a control $\{u(t) : t \in [0, T]\}$ with $T > 0$ that minimizes

$$J_0(u) = \|u\|_0 = \int_0^T |u(t)|^0 dt$$

subject to

$$x(T) = 0,$$

and

$$\|u\|_\infty \leq 1.$$

We call the solution of this optimal control problem the L^0-*optimal control*, or the *maximum hands-off control*.

The stage cost function (8.10) is discontinuous and non-convex, as shown in Figure 8.2. By borrowing the idea of sparse representation to use the ℓ^1 norm for

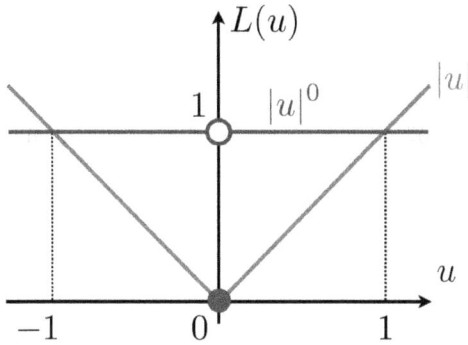

Figure 8.2. Stage cost functions $|u|^0$ and $|u|$ in L^0 and L^1 optimal control problems.

ℓ^0 norm optimization, we introduce the following cost function with the L^1 norm:

$$J_1(u) \triangleq \|u\|_1 = \int_0^T |u(t)|dt. \tag{8.12}$$

As shown in Figure 8.2, the stage cost function $|u|$ is an approximation of $|u|^0$. In fact, this approximation is mathematically explained as the *convex relaxation*. That is, the L^1 norm $\|u\|_1$ is the convex relaxation of $\|u\|_0$ when $\|u\|_\infty \leq 1$. See [112, Section 1.3.2] for details.

Now we formulate the L^1-*optimal control problem.*

L^1-optimal control problem (L^1-OPT)

For the linear time-invariant system

$$\dot{x}(t) = Ax(t) + bu(t), \quad t \geq 0, \quad x(0) = \xi \in \mathbb{R}^d,$$

find a control $\{u(t) : t \in [0, T]\}$ with $T > 0$ that minimizes

$$J_1(u) = \|u\|_1 = \int_0^T |u(t)|dt$$

subject to

$$x(T) = 0,$$

and

$$\|u\|_\infty \leq 1.$$

We call the solution of this optimal control problem the L^1-*optimal control.* This optimal control is also known as *minimum fuel control,* which was widely studied in 60s for rocket control.

The L^1-optimal control (L^1-OPT) is a convex optimization problem since the stage cost $L(u) = |u|$ is convex in u and the constraints are also convex. Although the variable u is a function, which is a member of infinite dimensional function space $L^\infty(0, T)$, the problem can be reduced to a finite-dimensional optimization problem via time discretization studied in Chapter 9.

8.4 L^1-optimal Control

Here we investigate properties of L^1 optimal control by using necessary conditions from Pontryagin's minimum principle.

For the L^1-optimal control problem (L^1-OPT), the Hamiltonian is given by

$$H^\eta(x, p, u) = p^\top (Ax + bu) + \eta|u|. \tag{8.13}$$

We first consider the case $\eta = 1$ (the normal case). Let u^* denote the L^1 optimal control, and x^* and p^* the associated optimal state and costate, respectively. From the minimum condition in the minimum principle, we have

$$u^*(t) = \arg\min_{|u|\leq 1} H^1(x^*(t), p^*(t), u)$$

$$= \arg\min_{|u|\leq 1} \left\{ p^*(t)^\top (Ax^*(t) + bu) + |u| \right\} \tag{8.14}$$

$$= \arg\min_{|u|\leq 1} \left\{ p^*(t)^\top bu + |u| \right\}$$

Now from

$$p^*(t)^\top bu + |u| = \begin{cases} \left(p^*(t)^\top b + 1\right)u, & \text{if } u \geq 0 \\ \left(p^*(t)^\top b - 1\right)u, & \text{if } u < 0 \end{cases} \tag{8.15}$$

we have the solution to the minimization problem in (8.14) as

$$u^*(t) = -\text{dez}\left(p^*(t)^\top b\right), \tag{8.16}$$

where $\text{dez}(\cdot)$ is called the *dead-zone function* defined by

$$\text{dez}(w) \triangleq \begin{cases} -1, & \text{if } w < -1 \\ 0, & \text{if } -1 < w < 1 \\ 1, & \text{if } 1 < w \end{cases}$$

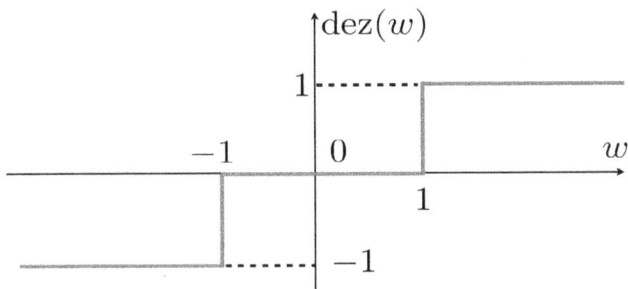

Figure 8.3. Dead-zone function dez(w).

$$\text{dez}(w) \in [-1, 0], \quad \text{if } w = -1$$
$$\text{dez}(w) \in [0, 1], \quad \text{if } w = 1 \tag{8.17}$$

Figure 8.3 shows the graph of the dead-zone function.

Exercise 8.2. Show that (8.16) is the solution to the minimization problem (8.14).

If there is a time interval (t_1, t_2) on which $p^*(t)^\top b \equiv \pm 1$ holds, then from (8.17), we cannot uniquely determine $u^*(t)$ on this interval. We call such a time interval a *singular interval*. If an L^1-optimal control problem has a singular interval whose length is positive, then we call the problem a *singular problem*. On the other hand, if

$$\mu\left(\{t \in [0, T] : |p^*(t)^\top b| = 1\}\right) = 0 \tag{8.18}$$

holds, then the L^1-optimal control problem is said to be *non-singular*. The following lemma gives a sufficient condition for the non-singularity.

Lemma 8.3. *If (A, b) in (8.9) is controllable and A is nonsingular, then (8.18) holds (i.e., the L^1-optimal control problem is non-singular).*

From now on, we say (A, b) *is non-singular* if (A, b) is controllable and A is non-singular.

Exercise 8.4. Prove Lemma 8.3.

From Lemma (8.3), if (A, b) is non-singular, then we have

$$p^*(t)^\top b \neq \pm 1, \quad \text{for almost all } t \in [0, T]. \tag{8.19}$$

Then, from (8.16) and (8.17), the L^1-optimal control takes values ± 1 or 0 for almost all t in $[0, T]$. We call such a control a *bang-off-bang* control. Figure 8.4 illustrates the bang-off-bang property of L^1-optimal control. We summarize this property as a theorem.

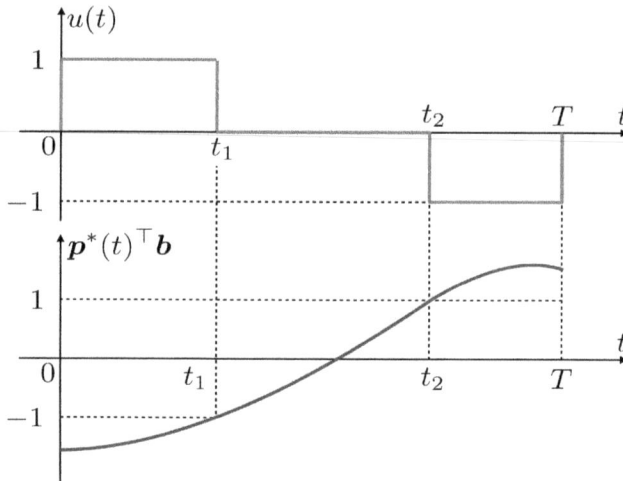

Figure 8.4. L^1-optimal control (bang-off-bang) $u^*(t)$ (top) and function $p^*(t)^\top b$.

Theorem 8.5. *Assume that (A, b) is non-singular. Then the L^1-optimal control is bang-off-bang (if it exists).*

The bang-off-bang property is important to examine the relation between L^1 and L^0 controls as shown in the next section.

Remark 8.6. *The function $p^*(t)^\top b$ is given by*

$$p^*(t)^\top b = \left(e^{-A^\top t}p^*(0)\right)^\top b = p^*(0)^\top e^{-At}b, \qquad (8.20)$$

from the adjoint equation for $p^*(t)$ in (7.41). Therefore the function $p^*(t)^\top b$ is continuous, and represented by

$$p^*(0)^\top e^{-At}b = \sum_{i=1}^{d} c_i(t)e^{-\lambda_i t}, \qquad (8.21)$$

where λ_i is the i-th eigenvalue of A and $c_i(t)$ is a polynomial with degrees up to d. It follows that the number of switching in $u^*(t)$ is finite, and the value changes between 1 and 0 or -1 and 0, and never changes between 1 and -1. Therefore, if $u^*(t)$ switches, then there exists a time duration with positive length on which $u^*(t) = 0$.

Finally, let us consider the case $\eta = 0$ (the abnormal case). The Hamiltonian is given by

$$H^0(x, p, u) = p^\top(Ax + bu). \qquad (8.22)$$

Then the optimal control $u^*(t)$ satisfies

$$
u^*(t) = \underset{u\in[-1,1]}{\arg\min}\ H^0(x^*(t), p^*(t), u)
$$

$$
= \underset{u\in[-1,1]}{\arg\min}\ p^*(t)^\top bu
$$

$$
= \begin{cases}
-1, & \text{if } p^*(t)^\top b > 0, \\
1, & \text{if } p^*(t)^\top b < 0, \\
[-1, 1], & \text{if } p^*(t)^\top b = 0.
\end{cases}
$$

(8.23)

If (A, b) is controllable, then $p^*(t)^\top b \neq 0$, and hence the control is *bang-bang*, taking values of ± 1. With this control, the L^1-optimal value is

$$
J_1(u^*) = \int_0^T |u^*(t)|dt = T. \tag{8.24}
$$

Moreover, if $T^*(\xi) < T < \infty$, then there exists the minimum-time control $u^*_{\text{time}}(t)$ to achieve $x(T^*(\xi)) = 0$. By using this minimum-time control, define the following control:

$$
\bar{u}(t) = \begin{cases}
u^*_{\text{time}}(t), & \text{if } 0 \leq t \leq T^*(\xi), \\
0, & \text{if } T^*(\xi) < t \leq T.
\end{cases} \tag{8.25}
$$

It is easily shown that $\bar{u}$ is a feasible control, that is $\bar{u} \in \mathcal{U}(T, \xi)$. Also, with this $\bar{u}$, we have

$$
J_1(\bar{u}) = \int_0^T |\bar{u}(t)|dt = \int_0^{T^*(\xi)} |u^*_{\text{time}}(t)|dt = T^*(\xi) < T = J_1(u^*).
$$

(8.26)

Hence, the control $u^*(t)$ can never be L^1 optimal, and hence the case $\eta = 0$ never happens.

The abnormal case ($\eta = 0$) happens when $T = T^*(\xi)$. In this case, the set of feasible controls is $U(T^*(\xi), \xi) = \{u^*_{\text{time}}\}$, a singleton of the minimum-time control, and hence the cost function is meaningless to choose a control from the feasible set. In this book, we do not discuss the abnormal case any further.

8.5 Equivalence Between L^0 and L^1 Optimal Controls

In this section, we study the equivalence between L^0 and L^1 optimal controls.
The following theorem is a fundamental theorem for the equivalence.

Theorem 8.7. *Assume that there exists an L^1-optimal control that is bang-off-bang. Then it is also L^0 optimal.*

Proof: Define $J_0(u) \triangleq \|u\|_0$ and $J_1(u) \triangleq \|u\|_1$. From the assumption, there exists an L^1-optimal control u_1^* that is bang-off-bang. Since u_1^* is a feasible control, the set of feasible controls $\mathcal{U}(T, \xi)$ is non-empty. Then, for any $u \in \mathcal{U}(T, \xi)$ we have

$$J_1(u) = \int_0^T |u(t)| dt = \int_{\text{supp}(u)} |u(t)| dt \leq \int_{\text{supp}(u)} 1 dt = J_0(u). \qquad (8.27)$$

Since u_1^* is bang-off-bang, we have

$$J_1(u_1^*) = \int_0^T |u_1^*(t)| dt = \int_{\text{supp}(u_1^*)} 1 dt = J_0(u_1^*). \qquad (8.28)$$

From (8.27) and (8.28), we have

$$J_0(u_1^*) \leq J_0(u), \quad \forall u \in \mathcal{U}(T, \xi), \qquad (8.29)$$

and hence u_1^* minimizes $J_0(u)$. That is, u_1^* is also L^0 optimal. □

From Theorem 8.5, if (A, b) is non-singular, then the L^1-optimal control is bang-off-bang, that is, the optimal control $u^*(t)$ takes values 0 or ± 1 for almost all $t \in [0, T]$. From this property, we can obtain the following theorem.

Theorem 8.8. *Assume that there exists at least one L^1-optimal control. Assume also that (A, b) is non-singular. Then there exists at least one L^0-optimal control, and the set of L^0-optimal controls is equivalent to the set of L^1-optimal controls.*

Proof: Let $\mathcal{U}_0^*$ and $\mathcal{U}_1^*$ be the sets of L^0 and L^1 optimal controls, respectively. From the assumption, $\mathcal{U}_1^*$ is non-empty. Take $u_1^* \in \mathcal{U}_1^*$ arbitrarily. Then, from Theorem 8.5, u_1^* is bang-off-bang. It follows from Theorem 8.7 that $u_1^* \in \mathcal{U}_0^*$, and hence $\mathcal{U}_1^* \subset \mathcal{U}_0^*$.

Then we prove $\mathcal{U}_0^* \subset \mathcal{U}_1^*$. Take $u_0^* \in \mathcal{U}_0^* \subset \mathcal{U}(T, \xi)$ arbitrarily. Take also $u_1^* \in \mathcal{U}_1^* \subset \mathcal{U}(T, \xi)$ independently. From (8.28) and the L^1 optimality of u_1^*, we have

$$J_0(u_1^*) = J_1(u_1^*) \leq J_1(u_0^*). \qquad (8.30)$$

On the other hand, from (8.27) and the L^0 optimality of u_0^*, we have

$$J_1(u_0^*) \leq J_0(u_0^*) \leq J_0(u_1^*). \qquad (8.31)$$

From (8.30) and (8.31), we have

$$J_0(u_1^*) = J_1(u_1^*) \leq J_1(u_0^*) \leq J_0(u_0^*) \leq J_0(u_1^*). \qquad (8.32)$$

It follows that $J_1(u_1^*) = J_1(u_0^*)$, and u_0^* minimizes $J_1(u)$. That is, we have $u_0^* \in \mathcal{U}_1^*$ and hence $\mathcal{U}_0^* \subset \mathcal{U}_1^*$. □

8.6   Existence of L^0-Optimal Control

Here we consider the existence of L^0-optimal control.

8.6.1 L^p-Optimal Control

To consider the existence of L^0-optimal control, we introduce the L^p-*optimal control* with $p \in (0, 1)$. The optimal control problem is described as follows:

L^p-optimal control problem (L^p-OPT)

For the linear time-invariant system

$$\dot{x}(t) = Ax(t) + bu(t), \quad t \geq 0, \quad x(0) = \xi \in \mathbb{R}^d,$$

find a control $\{u(t) : t \in [0, T]\}$ with $T > 0$ that minimizes

$$J_p(u) = \|u\|_p^p = \int_0^T |u(t)|^p dt$$

with $p \in (0, 1)$, subject to

$$x(T) = 0,$$

and

$$\|u\|_\infty \leq 1.$$

First, we prove an interesting relation between the L^p norm[2] with $p \in (0, 1)$ and the L^0 norm.

Lemma 8.9. *Suppose $u \in L^1(0, T)$. Then u is also in $L^p(0, T)$ for any $p \in (0, 1)$, and*

$$\lim_{p \to 0+} \|u\|_p^p = \|u\|_0. \tag{8.33}$$

Exercise 8.10. Prove Lemma 8.9.

---

2.   Strictly speaking, the L^p "norm" with $p \in (0, 1)$ is not a proper norm since the triangle inequality does not always hold.

Now, let us look into the L^p-optimal control with $p \in (0, 1)$. The Hamiltonian is given by

$$H^\eta(x, p, u) = p^\top(Ax + bu) + \eta|u|^p. \tag{8.34}$$

Let us consider the normal case ($\eta = 1$). From the minimum condition in Pontryagin's minimum principle, we have

$$
\begin{aligned}
u^*(t) &= \arg\min_{u \in [-1,1]} H^\eta(x^*(t), p^*(t), u) \\
&= \arg\min_{u \in [-1,1]} \left\{ p^*(t)^\top bu + |u|^p \right\} \\
&= \begin{cases}
-1, & \text{if } p^*(t)^\top b > 1 \\
0, & \text{if } -1 < p^*(t)^\top b < 1 \\
1, & \text{if } p^*(t)^\top b < -1 \\
\{-1, 0\}, & \text{if } p^*(t)^\top b = 1 \\
\{0, 1\}, & \text{if } p^*(t)^\top b = -1
\end{cases}
\end{aligned}
\tag{8.35}
$$

From this, L^p-optimal control is always *bang-off-bang*. We mention this in the following theorem.

Theorem 8.11. *The L^p-optimal control with $p \in (0, 1)$ is bang-off-bang (if it exists).*

Also, it is shown that the set of L^p-optimal control is equivalent to the set of L^0-optimal controls.

Theorem 8.12. *Assume that there exists at least one L^p-optimal control with $p \in (0, 1)$. Let $\mathcal{U}_0^*$ and $\mathcal{U}_p^*$ be the sets of L^0 and L^p optimal controls respectively. Then we have*

$$\mathcal{U}_0^* = \mathcal{U}_p^*. \tag{8.36}$$

Exercise 8.13. Prove Theorem 8.12.

From Theorems 8.11 and 8.12, we have the following theorem.

Theorem 8.14. *The L^0-optimal control is bang-off-bang (if it exists).*

The difference of Theorem 8.12 from Theorem 8.8 for the L^1-optimal control is that for the L^p optimal control we do not need the assumption of the non-singularity of (A, b). This is the key to prove the existence of L^0 optimal control.

8.6.2 Existence Theorems

From Theorem 8.12, if we show the existence of L^p-optimal control for some $p \in (0, 1)$, then $\mathcal{U}_0^*$ is non-empty, and hence there exists at least one L^0-optimal control. The following theorem is on the existence of L^p-optimal control with $p > 0$.[3]

Theorem 8.15. *Suppose that the initial state $\xi \in \mathbb{R}^d$ and the time $T > 0$ are chosen such that $\xi \in \mathcal{R}(T)$. Then, then there exists an L^p-optimal control with $p > 0$.*

Proof: Assume $\xi \in \mathbb{R}^d$ and $T > 0$ satisfy $\xi \in \mathcal{R}(T)$. Then there exists a feasible control $u \in \mathcal{U}(T, \xi)$, and hence the feasible set $\mathcal{U}(T, \xi)$ is non-empty. Define

$$J_p^* \triangleq \inf\{\|u\|_p^p : u \in \mathcal{U}(T, \xi)\}. \tag{8.37}$$

Since $u \in \mathcal{U}(T, \xi)$ satisfies $\|u\|_\infty \leq 1$, we have $J_p^* < \infty$. Then, from the definition of J_p^*, there exists a sequence $\{u_l\}_{l \in \mathbb{N}} \subset \mathcal{U}(T, \xi)$ such that

$$\lim_{l \to \infty} \|u_l\|_p^p = J_p^*. \tag{8.38}$$

Now, since $u_l \in \mathcal{U}(T, \xi)$, we have

$$\|u_l\|_\infty \leq 1, \tag{8.39}$$

and hence $\{u_l\}_{l \in \mathbb{N}} \subset \mathcal{B}_\infty \triangleq \{u \in L^\infty(0, T) : \|u\|_\infty \leq 1\}$. It is known that the unit ball $\mathcal{B}_\infty$ is sequentially compact in the weak* topology of $L^\infty(0, T)$ [74, Theorem A.9]. That is, there exists a subsequence $\{u_{l'}\}_{l' \in S}$, $S \subset \mathbb{N}$, such that there exists $u_\infty \in \mathcal{B}_\infty$ and

$$\lim_{l' \to \infty} \int_0^T f(t)\big(u_{l'}(t) - u_\infty(t)\big)dt = 0, \tag{8.40}$$

for any $f \in L^1(0, T)$. Now, since $u_{l'} \in \mathcal{U}(T, \xi)$, we have

$$\xi = -\int_0^T e^{-At} b u_{l'}(t)dt, \quad \forall l' \in S. \tag{8.41}$$

On the other hand, from (8.40) with $f(t) = e^{-At}b$, we have

$$\lim_{l' \to \infty} \int_0^T e^{-At} b u_{l'}(t)dt = \int_0^T e^{-At} b u_\infty(t)dt. \tag{8.42}$$

---

3. To show the existence of L^0-optimal control, we just need to prove the existence of L^p-optimal control with $p \in (0, 1)$. However, Theorem 8.15 gives more general result.

That is,

$$\xi = -\int_0^T e^{-At} b u_\infty(t) dt. \tag{8.43}$$

Also since $u_\infty \in \mathcal{B}_\infty$, we have $\|u_\infty\|_\infty \le 1$. Therefore, $u_\infty \in \mathcal{U}(T, \xi)$.

Next, let us define

$$\theta_{l'} \triangleq \int_0^T u_{l'}(t)^p \operatorname{sgn}\big(u_\infty(t)^p\big) dt. \tag{8.44}$$

Similar to (8.40), for the sequence $\{u_l^p\}_{l\in\mathbb{N}} \subset \mathcal{B}_\infty$ and $f(t) = \operatorname{sgn}\big(u_\infty(t)^p\big)$, we have

$$\lim_{l'\to\infty} \theta_{l'} = \int_0^T u_\infty(t)^p \operatorname{sgn}\big(u_\infty(t)^p\big) dt = \int_0^T |u_\infty(t)|^p dt = \|u_\infty\|_p^p. \tag{8.45}$$

On the other hand, we have

$$|\theta_{l'}| \le \int_0^T |u_{l'}(t)|^p dt = \|u_{l'}\|_p^p \to J_p^*, \tag{8.46}$$

as $l' \to \infty$, and hence from (8.45), we have

$$\|u_\infty\|_\infty \le J_p^*. \tag{8.47}$$

Now, from (8.37), J_p^* is the minimum value of $\|u\|_p^p$ over $\mathcal{U}(T, \xi)$, and hence

$$\|u_\infty\|_p^p = J_p^*. \tag{8.48}$$

That is, $u_\infty \in \mathcal{U}(T, \xi)$ is an L^p-optimal control. $\square$

From Theorem 8.15 and Theorem 8.12, we have the following theorem.

Theorem 8.16 (Existence of L^0 optimal control). *If $\xi \in \mathcal{R}(T)$, then there exists an L^0-optimal control.*

The condition of $\xi \in \mathcal{R}(T)$ is equivalent to $\xi \in \mathcal{R}$ and $T \ge T^*(\xi)$. Hence, we have the following lemma.

Lemma 8.17. *Let u^* be the L^0-optimal control with $\xi \in \mathcal{R}(T)$. Then $\|u^*\|_0 \le T^*(\xi)$.*

Proof: This is easily shown by considering a feasible control in (8.25). $\square$

8.7   Maximum Hands-off Control of Rocket

Here we compute the maximum hands-off control of the rocket considered in Example 7.2 (p. 132) in the previous chapter. We assume the mass $m = 1$ for simplicity.

We now compute the L^1-optimal control. From (8.13), the Hamiltonian with $\eta = 1$ is given by

$$H(x, p, u) = p^\top \left(\begin{bmatrix} 0 & 1 \\ 0 & 0 \end{bmatrix} x + \begin{bmatrix} 0 \\ 1 \end{bmatrix} u \right) + |u| = p_1 x_1 + p_2 u + |u|, \quad (8.49)$$

where $p = (p_1, p_2)$. Let denote by u^* the L^1-optimal control, and $x^* = (x_1^*, x_2^*)$, $p^* = (p_1^*, p_2^*)$ the associated optimal state and costate, respectively. From (8.16), the L^1-optimal control $u^*(t)$ satisfies

$$u^*(t) = -\text{dez}\big(p_2^*(t)\big), \quad (8.50)$$

where dez$(\cdot)$ is the dead-zone function defined in (8.17) (see also Figure 8.3).

Then, the adjoint equation of the costate $p^*(t)$ becomes

$$\begin{bmatrix} \dot{p}_1^*(t) \\ \dot{p}_2^*(t) \end{bmatrix} = - \begin{bmatrix} 0 & 1 \\ 0 & 0 \end{bmatrix}^\top \begin{bmatrix} p_1^*(t) \\ p_2^*(t) \end{bmatrix} = \begin{bmatrix} 0 \\ -p_1^*(t) \end{bmatrix}. \quad (8.51)$$

The solution of this differential equation is given by

$$p_1^*(t) = \pi_1, \quad p_2^*(t) = \pi_2 - \pi_1 t, \quad (8.52)$$

where

$$\pi_1 = p_1^*(0), \quad \pi_2 = p_2^*(0). \quad (8.53)$$

It follows from (8.52) that if $\pi_1 \neq 0$ then $p_2^*(t)$ is a first-order linear function of t and hence $p_2^*(t)$ is monotone. Therefore, from (8.50) and the definition of the dead-zone function (8.17), switching occurs at most twice, and the value changes between -1 and 0, or between 0 and 1. From this observation, the L^1-optimal control is given as follows (for details, refer to [2, Section 8.5]).

Define the following regions (see Figure 8.5).

$$\gamma = \Big\{ (x_1, x_2) \in \mathbb{R}^2 : x_1 = -x_2|x_2|/2 \Big\}$$

$$R_1 = \Big\{ (x_1, x_2) \in \mathbb{R}^2 : x_1 > -x_2^2/2, \ x_2 \geq 0 \Big\}$$

$$R_2 = \Big\{ (x_1, x_2) \in \mathbb{R}^2 : x_1 < -x_2^2/2, \ x_2 > 0 \Big\}$$

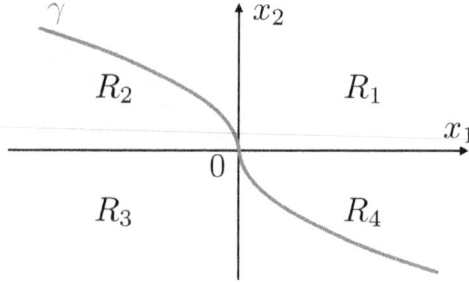

Figure 8.5. Curve γ (thick solid line) and regions R_1, R_2, R_3, and R_4.

$$R_3 = \left\{ (x_1, x_2) \in \mathbb{R}^2 : x_1 < x_2^2/2, \; x_2 \le 0 \right\}$$
$$R_4 = \left\{ (x_1, x_2) \in \mathbb{R}^2 : x_1 > x_2^2/2, \; x_2 < 0 \right\} \tag{8.54}$$

Also, define the following two regions:

$$V_- = \left\{ (x_1, x_2) \in \mathbb{R}^2 : -x_2/2 - x_1/x_2 \ge T \right\}$$
$$V_+ = \left\{ (x_1, x_2) \in \mathbb{R}^2 : x_2/2 - x_1/x_2 \ge T \right\} \tag{8.55}$$

Then, the L^1-optimal control is given as follows:

1. If $(\xi_1, \xi_2) \in R_1$ or $(\xi_1, \xi_2) \in R_4 \cap V_-$, then the optimal control is given by

$$u^*(t) = \begin{cases} -1, & \text{if } 0 \le t < t_1 \\ 0, & \text{if } t_1 \le t < t_2 \\ 1, & \text{if } t_2 \le t \le T \end{cases} \tag{8.56}$$

where

$$t_1 = \frac{T + \xi_2 - \sqrt{(T - \xi_2)^2 - 4\xi_1 - 2\xi_2^2}}{2},$$
$$t_2 = \frac{T + \xi_2 + \sqrt{(T - \xi_2)^2 - 4\xi_1 - 2\xi_2^2}}{2}. \tag{8.57}$$

2. If $(\xi_1, \xi_2) \in R_3$ or $(\xi_1, \xi_2) \in R_2 \cap V_+$ then the optimal control is given by

$$u^*(t) = \begin{cases} 1, & \text{if } 0 \le t < t_3 \\ 0, & \text{if } t_3 \le t < t_4 \\ -1, & \text{if } t_4 \le t \le T \end{cases} \tag{8.58}$$

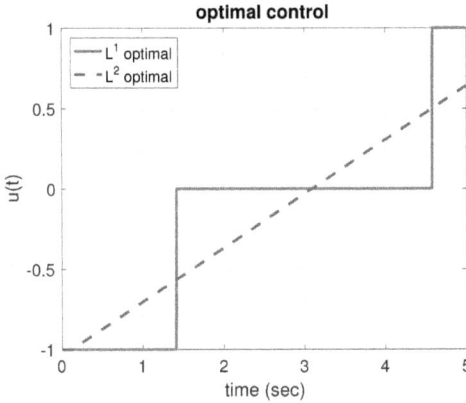

Figure 8.6. L^1-optimal control (solid line) and L^2-optimal control (dashed line).

where

$$t_3 = \frac{T - \xi_2 - \sqrt{(T + \xi_2)^2 + 4\xi_1 - 2\xi_2^2}}{2},$$

$$t_4 = \frac{T - \xi_2 + \sqrt{(T + \xi_2)^2 + 4\xi_1 - 2\xi_2^2}}{2}. \tag{8.59}$$

3. If $(\xi_1, \xi_2) \in \gamma$ then the optimal control is given by

$$u^*(t) = \begin{cases} -\mathrm{sgn}(\xi_2), & \text{if } 0 \le t < |\xi_2| \\ 0, & \text{if } |\xi_2| \le t \le T \end{cases} \tag{8.60}$$

4. If $(\xi_1, \xi_2) \in R_4 \cap (V_-)^c$ or $(\xi_1, \xi_2) \in R_2 \cap (V_+)^c$,[4] then the control problem is *singular*, and the optimal control cannot be uniquely determined from the minimum principle.

Figure 8.6 shows the L^1-optimal control with the final time $T = 5$ and the initial state $(\xi_1, \xi_2) = (1, 1) \in R_1$. Figure 8.7 shows the associated optimal state trajectory $\{(x_1^*(t), x_2^*(t)) : 0 \le t \le 5\}$. In these figures, we also show the results of L^2-optimal control that minimizes the L^2 cost function

$$\|u\|_2^2 = \int_0^T |u(t)|^2 dt, \tag{8.61}$$

among the feasible controls. From Figure 8.6, we can see that the L^1-optimal control is sparse while the L^2-optimal control is not. In fact, the L^1-optimal control

---

4. $(\cdot)^c$ denotes the complement set.

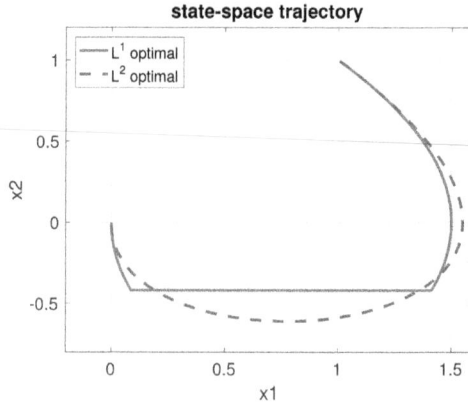

Figure 8.7. Optimal state trajectory $(x_1^*(t), x_2^*(t))$: L^1-optimal control (solid line) and L^2-optimal control (dashed line).

is bang-off-bang, and hence this is equivalent to L^0-optimal control. That is, the L^1-optimal control has the maximum length of time duration on which the control is exactly zero. From (8.57), this time length is given by

$$[t_1, t_2] = [3 - \sqrt{10}/2, 3 + \sqrt{10}/2] \approx [1.4189, 4.5811]. \qquad (8.62)$$

and the L^0 norm of the L^1-optimal control u^* is $\|u^*\|_0 = \sqrt{10} \approx 3.1623$. On this time duration, the state trajectory $(x_1^*(t), x_2^*(t))$ is parallel to the x_1 axes. Since x_1 is the portion and x_2 is the velocity of the rocket, this state trajectory means that the rocket moves at a constant velocity. The rocket consumes no fuel on this time duration, and hence we can cut fuel consumptions and also we can reduce CO2 emissions, etc. That is, the control is *green*. It is clear that L^2-optimal control does not have such a nice property of sparsity.

8.8   Further Reading

For the L^1-optimal control (minimum fuel control), the most detailed information can be obtained from the classical book by Athans and Falb [2]. The equivalence theorem between L^0 and L^1 optimal controls was first proved in [82, 83].

   In this book, we consider only linear systems, but the equivalence holds for nonlinear systems of the following type:

$$\dot{x}(t) = f(x(t)) + g(x(t))u(t), \quad t \geq 0. \qquad (8.63)$$

See [82, 83] for details.

Necessary conditions of the L^0-optimal control are also obtained in [19] by the non-smooth version of Pontryagin's minimum (or maximum) principle [24]. For the theory of L^p spaces, see [65, 99, 114].

For feedback control implementation of maximum hands-off control, one can adopt the model predictive control [55, 86] and the self-triggered control [83].

An interesting extension of maximum hands-off control is distributed control for multi-agent systems discussed in [52].

DOI: 10.1561/9781680837254.ch9

Chapter 9

Numerical Optimization by Time Discretization

As we have seen in Section 8.7, the maximum hands-off control (or L^1-optimal control) is obtained in a closed form when the plant is very simple as the double integrator given in Example 7.2 (p. 132). However, for a general system

$$\dot{x}(t) = Ax(t) + bu(t), \quad t \geq 0, \quad x(0) = \xi \in \mathbb{R}^d, \tag{9.1}$$

we need to rely on numerical computation to obtain the optimal control. In this chapter, we introduce the method of time discretization to numerically obtain the L^1-optimal control.

┌─ Key ideas of Chapter 9 ─────────────────────────────────────

- By time discretization, the L^1-optimal control problem (L^1-OPT) is reduced to a finite-dimensional ℓ^1 optimization problem.
- In time discretization, the control is assumed to be piecewise constant by a zero-order hold.
- The reduced ℓ^1 optimization can be efficiently solved by ADMM.

└──

9.1 Time Discretization

First, we discretize the time interval $[0, T]$ into n subintervals as

$$[0, T] = [0, h) \cup [h, 2h) \cup \cdots \cup [nh - h, nh], \tag{9.2}$$

where $h > 0$ is the sampling time and $n \in \mathbb{N}$ is the number of subintervals such that $T = nh$.

On each subinterval, we assume the control $u(t)$ is constant. More precisely, we assume the control is given by

$$u(t) = u(kh) = u_\mathrm{d}[k], \quad t \in [kh, (k + 1)h), \quad k = 0, 1, 2, \ldots, n - 1. \tag{9.3}$$

This is the output of a *zero-order hold* of a discrete-time signal

$$u_\mathrm{d} \triangleq \{u_\mathrm{d}[0], u_\mathrm{d}[1], \ldots, u_\mathrm{d}[n - 1]\}. \tag{9.4}$$

This assumption is actually reasonable for networked digital control systems where control values are computed in a digital computer, transmitted through a wireless communication network, and applied to an actuator through a D/A converter. The zero-order hold is the simplest model of a D/A converter.

Let us compute the state transition under the zero-order assumption on the control. The solution to the state-space equation in (9.1) is given by (see Exercise 7.1 on p. 132)

$$x(t_1) = e^{A(t_1 - t_0)} x(t_0) + \int_{t_0}^{t_1} e^{A(t_1 - \tau)} bu(\tau) d\tau, \tag{9.5}$$

where $0 \leq t_0 \leq t_1$. Take

$$t_0 = kh, \quad t_1 = kh + h, \quad k \in \{0, 1, 2, \ldots, n - 1\}. \tag{9.6}$$

Then from (9.5) we have

$$x(kh + h) = e^{Ah} x(kh) + \int_{kh}^{kh+h} e^{A(kh+h-\tau)} bu(\tau) d\tau$$

$$= e^{Ah} x(kh) + \int_0^h e^{A(h-t)} bu(t + kh) dt. \tag{9.7}$$

Define

$$x_\mathrm{d}[k] \triangleq x(kh), \quad u_\mathrm{d}[k] \triangleq u(kh), \quad k = 0, 1, \ldots, n - 1, \tag{9.8}$$

and

$$x_d[n] \triangleq x(T).$$

(9.9)

From the zero-order-hold assumption (9.3), the control $u(t)$ takes a constant value $u_d[k] = u(kh)$ on the subinterval $[kh, kh+h)$ as shown in Figure 9.1. Then from (9.7) we have

$$x_d[k+1] = e^{Ah}x_d[k] + \left(\int_0^h e^{A(h-t)}b\,dt\right)u_d[k].$$

(9.10)

It follows that the differential equation (9.1) is transformed into the following difference equation:

$$x_d[k+1] = A_d x_d[k] + b_d u_d[k], \quad k = 0, 1, \ldots, n-1,$$

(9.11)

where

$$A_d \triangleq e^{Ah}, \quad b_d \triangleq \int_0^h e^{At}b\,dt.$$

(9.12)

Next, define the control vector

$$u \triangleq \begin{bmatrix} u_d[0] \\ u_d[1] \\ \vdots \\ u_d[n-1] \end{bmatrix} \in \mathbb{R}^n.$$

(9.13)

By using this, the terminal state $x(T)$ is described as

$$x(T) = x_d[n] = -\zeta + \Phi u,$$

(9.14)

where

$$\Phi \triangleq \begin{bmatrix} A_d^{n-1}b_d & A_d^{n-2}b_d & \cdots & b_d \end{bmatrix}, \quad \zeta \triangleq -A_d^n \xi.$$

(9.15)

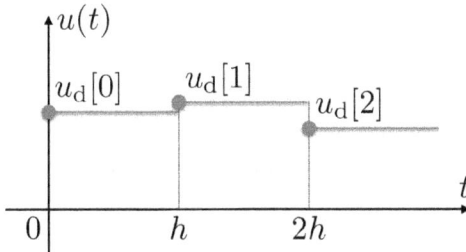

Figure 9.1. Zero-order hold output of discrete-time signal $\{u_d[k]\}$.

Exercise 9.1. Show the equation (9.14) by solving the difference equation (9.11).

9.2 Controllability of Discretized Systems

The discrete-time system (9.11) is called the *zero-order-hold discretization* or *step-invariant discretization* of the continuous-time system (9.1). Figure 9.2 shows the zero-order-hold discretization of (9.1). In this figure $\mathcal{H}_h$ is the *zero-order hold* with sampling time h, which outputs a constant value $u_d[k] = u(kh)$ over $[kh, (k + 1)h)$, $k = 0, 1, 2, \ldots$ (see Figure 9.1). Also, $\mathcal{S}_h$ is the *ideal sampler* that outputs the sampled value $x_d[k] = x(kh)$, $k = 0, 1, 2, \ldots$ of the continuous-time signal $x(t)$.

As discussed above, the discrete-time system from $u_d[k]$ to $x_d[k]$ in Figure 9.2 is a linear time-invariant discrete-time system as in (9.11). Then, under this discretization, the stability is preserved; if A is stable, that is, if the eigenvalues of A have non-positive real parts then A_d is Schur stable, that is, the eigenvalues of A_d lie in the closed unit circle in $\mathbb{C}$. This is easily shown from the spectral mapping theorem: the set of the eigenvalues of $A_d = e^{Ah}$ is given by $\{e^{\lambda_1 h}, \ldots, e^{\lambda_d h}\}$, where λ_i is the i-th eigenvalue of A.

On the other hand, we cannot say the controllability is not always preserved under the zero-order-hold discretization. To discuss this, we introduce the concept of *pathological sampling*.

Definition 9.2 (pathological sampling). *Let $\lambda(A)$ be the set of eigenvalues of A. The sampling time $h > 0$ is said to be* pathological *if there exist $\lambda_1, \lambda_2 \in \lambda(A)$ such that*

1. $\lambda_1 \neq \lambda_2$
2. $\operatorname{Re} \lambda_1 = \operatorname{Re} \lambda_2$
3. *there exists $k \in \{\pm 1, \pm 2, \ldots\}$ such that*

$$\operatorname{Im} \lambda_1 - \operatorname{Im} \lambda_2 = \frac{2\pi k}{h}. \qquad (9.16)$$

Intuitively, pathological sampling synchronizes an oscillation mode in the plant. The following illustrates pathological sampling.

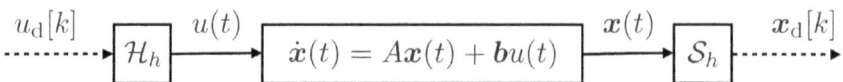

Figure 9.2. Zero-order-hold discretization: continuous-time system $\dot{x}(t) = Ax(t) + bu(t)$ is discretized by zero-order hold $\mathcal{H}_h$ and ideal sampler $\mathcal{S}_h$ with sampling time h.

Example 9.3. *Let us consider a linear system*

$$\ddot{y}(t) = -y(t), \quad y(0) = 0, \quad \dot{y}(0) = 1. \tag{9.17}$$

Then the solution of this differential equation is given by

$$y(t) = \sin t. \tag{9.18}$$

If we sample this output with sampling period $h = \pi$, then we have

$$y(kh) = \sin kh = 0, \quad k = 0, 1, 2, \ldots \tag{9.19}$$

This is an example of pathological sampling. The state-space representation of (9.17) *is given by*

$$\frac{d}{dt}\begin{bmatrix} x_1(t) \\ x_2(t) \end{bmatrix} = \begin{bmatrix} 0 & 1 \\ -1 & 0 \end{bmatrix}\begin{bmatrix} x_1(t) \\ x_2(t) \end{bmatrix}, \quad \begin{bmatrix} x_1(0) \\ x_2(0) \end{bmatrix} = \begin{bmatrix} 0 \\ 1 \end{bmatrix}, \tag{9.20}$$

where $x_1(t) \triangleq y(t)$ and $x_2(t) \triangleq \dot{y}(t)$. Then the matrix

$$A = \begin{bmatrix} 0 & 1 \\ -1 & 0 \end{bmatrix} \tag{9.21}$$

has two eigenvalues $\lambda_{\pm} = \pm j$ satisfying

$$\operatorname{Im} \lambda_+ - \operatorname{Im} \lambda_- = 2 = \frac{2\pi k}{h}, \tag{9.22}$$

with $k = 1$. Therefore, $h = \pi$ is certainly pathological.

When the sampling is non-pathological, then the controllability is preserved as shown in the following theorem.

Theorem 9.4. *Assume that the sampling time h is non-pathological. Then (A, b) is controllable if and only if (A_d, b_d) is controllable.*

The proof is found in [22].

9.3 Reduction to Finite-dimensional Optimization

Now we reduce the L^1-optimal control problem (L^1-OPT) (p. 155) into a finite-dimensional ℓ^1 optimization problem by the time discretization.

First, the constraint on the magnitude of control $\|u\|_\infty \leq 1$ is equivalently written by

$$|u_d[k]| \leq 1, \quad \forall k \in \{0, 1, 2, \ldots, n-1\}, \tag{9.23}$$

under the zero-order-hold assumption (9.3). Let us denote by $\|u\|_{\ell^\infty}$ the ℓ^∞ norm of a vector u (see (2.29) in Chapter 2). Then the above inequality is equivalent to

$$\|u\|_{\ell^\infty} \le 1. \tag{9.24}$$

Next, under the zero-order-hold assumption, the L^1 cost function becomes

$$J_1(u) = \int_0^T |u(t)|dt$$

$$= \sum_{k=0}^{n-1} \int_{kh}^{(k+1)h} |u(t)|dt$$

$$= \sum_{k=0}^{n-1} \int_{kh}^{(k+1)h} |u_\mathrm{d}[k]|dt \tag{9.25}$$

$$= \sum_{k=0}^{n-1} |u_\mathrm{d}[k]|h$$

$$= h\|u\|_{\ell^1}.$$

Now the L^1-optimal control problem (L^1 OPT) is reduced to the following finite-dimensional ℓ^1 optimization problem:

$$\underset{u \in \mathbb{R}^n}{\text{minimize}}\ \|u\|_{\ell^1}\ \text{subject to}\ \ \Phi u = \zeta,\ \|u\|_{\ell^\infty} \le 1. \tag{9.26}$$

This optimization problem is a *convex* optimization since the cost function (ℓ^1 norm) is a convex function, and the constraint set

$$\mathcal{C} \triangleq \{u \in \mathbb{R}^n : \Phi u = \zeta,\ \|u\|_{\ell^\infty} \le 1\} \tag{9.27}$$

is a convex set in $\mathbb{R}^n$. We can easily solve this problem by using CVX in MATLAB (see Section 3.3 in Chapter 3, p. 53). A MATLAB program to solve the ℓ^1 optimization (9.26) using CVX is given in Section 9.5.

9.4 Fast Algorithm by ADMM

If the order d of the system and the number n for discretization are not so large, you can obtain a solution easily by CVX. However, if you want to use the control in a feedback loop, then you must solve the problem in *real time*. Also, in real systems, the control algorithm should be implemented in a microcomputer, which often has just a cheap computational ability and is hard to run CVX. In such a case, we need to implement a fast and simple algorithm for the specific ℓ^1 optimization problem (9.26). For this purpose, we can use the efficient algorithms studied in Chapter 4. In particular, we here use ADMM (Alternating Direction Method of Multipliers) studied in Section 4.5 to solve (9.26).

First, define the unit ball $C_1 \subset \mathbb{R}^n$ with the ℓ^∞ norm by

$$C_1 \triangleq \{u \in \mathbb{R}^n : \|u\|_{\ell^\infty} \leq 1\}. \tag{9.28}$$

Also, let C_2 be a singleton of $\zeta \in \mathbb{R}^d$, that is,

$$C_2 \triangleq \{\zeta\}. \tag{9.29}$$

Define the indicator functions of the sets C_1 and C_2, respectively, by

$$I_{C_1}(u) \triangleq \begin{cases} 0, & \text{if } \|u\|_{\ell^\infty} \leq 1, \\ \infty, & \text{if } \|u\|_{\ell^\infty} > 1, \end{cases} \tag{9.30}$$

$$I_{C_2}(x) \triangleq \begin{cases} 0, & \text{if } x = \zeta, \\ \infty, & \text{if } x \neq \zeta. \end{cases} \tag{9.31}$$

Then the optimization problem (9.26) is equivalently described by

$$\underset{u \in \mathbb{R}^n}{\text{minimize}} \ \{\|u\|_{\ell^1} + I_{C_1}(u) + I_{C_2}(\Phi u)\}. \tag{9.32}$$

Next, define new variables $z_0, z_1 \in \mathbb{R}^n$, $z_2 \in \mathbb{R}^d$ by

$$z_0 = z_1 = u, \quad z_2 = \Phi u. \tag{9.33}$$

Then the problem (9.32) becomes

$$\underset{u \in \mathbb{R}^n, z \in \mathbb{R}^\nu}{\text{minimize}} \ \{\|z_0\|_{\ell^1} + I_{C_1}(z_1) + I_{C_2}(z_2)\} \ \text{subject to} \ z = \Psi u, \tag{9.34}$$

where $\nu \triangleq 2n + d$, and

$$z \triangleq \begin{bmatrix} z_0 \\ z_1 \\ z_2 \end{bmatrix} \in \mathbb{R}^\nu, \quad \Psi \triangleq \begin{bmatrix} I \\ I \\ \Phi \end{bmatrix} \in \mathbb{R}^{\nu \times n}. \tag{9.35}$$

Defining two functions f_1 and f_2 by

$$f_1(u) \triangleq 0, \quad f_2(z) \triangleq \|z_0\|_{\ell^1} + I_{C_1}(z_1) + I_{C_2}(z_2) \tag{9.36}$$

we finally obtain the standard optimization problem for ADMM (see (4.96), p. 81):

$$\underset{u\in\mathbb{R}^n,\, z\in\mathbb{R}^\nu}{\text{minimize}}\ \ f_1(u) + f_2(z)\ \text{ subject to }\ z = \Psi u, \qquad (9.37)$$

for which ADMM algorithm is given by (see Section 4.5.1, p. 81)

$$u[k+1] := \underset{u\in\mathbb{R}^n}{\arg\min}\left\{ f_1(u) + \frac{1}{2\gamma}\left\|\Psi u - z[k] + v[k]\right\|_{\ell^2}^2 \right\}, \qquad (9.38)$$

$$z[k+1] := \text{prox}_{\gamma f_2}\big(\Psi u[k+1] + v[k]\big), \qquad (9.39)$$

$$v[k+1] := v[k] + \Psi u[k+1] - z[k+1]. \qquad (9.40)$$

Let us compute the functions in (9.38)–(9.40). First, since $f_1 = 0$, the first step (9.38) is minimization of a quadratic function, and it is reduced to the following linear equation:

$$\begin{aligned}
u[k+1] &= \underset{u\in\mathbb{R}^n}{\arg\min}\left\{ \frac{1}{2\gamma}\left\|\Psi u - z[k] + v[k]\right\|_{\ell^2}^2 \right\} \\
&= (\Psi^\top\Psi)^{-1}\Psi^\top(z[k] - v[k]).
\end{aligned} \qquad (9.41)$$

Note that $\Psi^\top\Psi = 2I + \Phi^\top\Phi$ is non-singular and the matrix

$$M \triangleq (\Psi^\top\Psi)^{-1}\Psi^\top \qquad (9.42)$$

can be computed off-line (i.e. outside the iteration).

The size of $\Psi^\top\Psi$ is $n \times n$, and if the number n of time discretization is very large, then the computation of the inversion may take large computational time. In this case, we can adopt the *matrix inversion lemma*

$$(X + UYV)^{-1} = X^{-1} - X^{-1}U(Y^{-1} + VX^{-1}U)^{-1}VX^{-1}. \qquad (9.43)$$

By this, the inverse matrix $(\Psi^\top\Psi)^{-1}$ can be rewritten as

$$(\Psi^\top\Psi)^{-1} = (2I + \Phi^\top\Phi)^{-1} = \frac{1}{2}I - \frac{1}{2}\Phi^\top(2I + \Phi\Phi^\top)^{-1}\Phi. \qquad (9.44)$$

This requires inversion of matrix $2I + \Phi\Phi^\top$ of size $d \times d$, and if $d \ll n$ then the computational time can be significantly reduced.

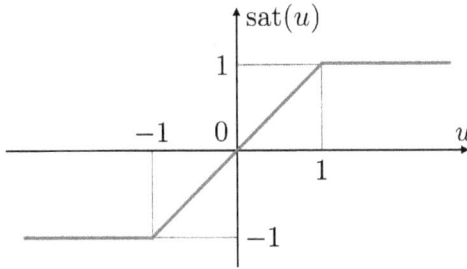

Figure 9.3. Saturation function $\mathrm{sat}(u) = \mathrm{sgn}(u)\min\{|u|, 1\}$.

The second step (9.39) in ADMM algorithm can be split into three simple optimization problems with variables z_0, z_1, and z_2 defined in (9.35). For the variable z_0, we use the proximal operator of the ℓ^1, which is the *soft-thresholding operator* defined in (4.46) (see also Figure 4.8 on p .70), that is, the i-th element of $\mathrm{prox}_{\gamma\|\cdot\|_{\ell^1}}(u)$ is given by

$$
\left[\mathrm{prox}_{\gamma\|\cdot\|_{\ell^1}}(u)\right]_i = [S_\gamma(u)]_i \triangleq
\begin{cases}
u_i - \gamma, & u_i \geq \gamma, \\
0, & |u_i| < \gamma, \\
u_i + \gamma, & u_i \leq -\gamma,
\end{cases}
\tag{9.45}
$$

where u_i is the i-th element of vector u.

For the variables z_1 and z_2, we need to compute the proximal operators of indicator functions. From (4.38) (p. 68), the proximal operator of the indicator function on a closed and convex set $\mathcal{C}$ is given by the projection $\Pi_{\mathcal{C}}$ onto $\mathcal{C}$. Therefore, the second step for variables z_1 and z_2 are reduced to $\Pi_{\mathcal{C}_1}$ and $\Pi_{\mathcal{C}_2}$.

The projection $\Pi_{\mathcal{C}_1}$ is given by

$$
\Pi_{\mathcal{C}_1}(u) =
\begin{bmatrix}
\mathrm{sat}(u_1) \\
\mathrm{sat}(u_2) \\
\vdots \\
\mathrm{sat}(u_n)
\end{bmatrix},
\qquad
\mathrm{sat}(u) \triangleq \mathrm{sgn}(u)\min\{|u|, 1\},
\tag{9.46}
$$

where the function sat($\cdot$) is called the *saturation function*. Figure 9.3 shows the graph of the saturation function. The other projection $\Pi_{\mathcal{C}_2} = \Pi_{\{\zeta\}}$ is simply given by

$$\Pi_{\mathcal{C}_2}(z) \triangleq \zeta. \tag{9.47}$$

In summary, the second step for variable z is given by

$$z[k+1] = \begin{bmatrix} S_\gamma\left(u[k+1] + v_0[k]\right) \\ \Pi_{\mathcal{C}_1}\left(u[k+1] + v_1[k]\right) \\ \zeta \end{bmatrix}, \tag{9.48}$$

where we split the vector $v[k]$ as $v = [v_0^\top, v_1^\top, v_2^\top]^\top$ consistent with the split of z in (9.35).

Now we obtain ADMM algorithm to solve the ℓ^1 optimization (9.26):

ADMM algorithm to solve the ℓ^1 optimization problem (9.26)

Initialization: give initial vectors $z[0]$, $v[0] \in \mathbb{R}^\nu$, and real number $\gamma > 0$
Iteration: for $k = 0, 1, 2, \ldots$ do

$$u[k+1] = M(z[k] - v[k]) \tag{9.49}$$

$$z[k+1] = \begin{bmatrix} S_\gamma\left(u[k+1] + v_0[k]\right) \\ \Pi_{\mathcal{C}_1}\left(u[k+1] + v_1[k]\right) \\ \zeta \end{bmatrix} \tag{9.50}$$

$$v[k+1] = v[k] + \Psi u[k+1] - z[k+1], \quad k = 0, 1, 2, \ldots \tag{9.51}$$

In this algorithm, the matrix M in (9.49) is given by (9.42).

As mentioned in [12], ADMM algorithm is very fast and needs just some dozens of iterations to obtain a solution with a sufficient precision. This property is very important if you adapt the finite-horizon L^1 optimal control to the *model predictive control* [70], where real-time computation is essential.

9.5   MATLAB Programs

We show MATLAB programs to solve the ℓ^1 optimization problem (9.26). One is a program using CVX. The other is an implementation of ADMM algorithm.

MATLAB program to solve ℓ^1 optimization problem (9.26) via CVX

```
clear
%% System model
% Plant matrices
A = [0,1;0,0];
b = [0;1];
d = length(b); %system size
% initial states
x0 = [1;1];
% Horizon length
T = 5;
%% Time discretization
% Discretization size
n = 10000; % grid size
h = T/n; % discretization interval
% System discretization
[Ad,bd] = c2d(A,b,h);
% Matrix Phi
Phi = zeros(d,n);
v = bd;
Phi(:,end) = v;
for j = 1:n-1
    v = Ad*v;
    Phi(:,end-j) = v;
end
% Vector zeta
zeta = -Ad^n*x0;
%% Convex optimization via CVX
cvx_begin
 variable u(n)
 minimize norm(u,1)
 subject to
   Phi*u == zeta;
   norm(u,inf) <= 1;
cvx_end
%% Plot
figure;
plot(0:T/n:T-T/n,u);
title('Sparse control');
```

MATLAB program to solve (9.26) with ADMM

```
clear
%% System model
% Plant matrices
A = [0,1;0,0];
b = [0;1];
d = length(b); %system size
% initial states
x0 = [1;1];
% Horizon length
T = 5;
%% Time discretization
% Discretization size
n = 1000; % grid size
h = T/n; % discretization interval
% System discretization
[Ad,bd] = c2d(A,b,h);
% Matrix Phi
Phi = zeros(d,n);
v = bd;
Phi(:,end) = v;
for j = 1:n-1
    v = Ad*v;
    Phi(:,end-j) = v;
end
% Vector zeta
zeta = -Ad^n*x0;
%% Convex optimization via ADMM
mu = 2*n+d;
Psi = [eye(n);eye(n);Phi];
M = (0.5*eye(n) - 0.5*Phi'*inv(2*eye(d)+Phi*Phi')*Phi)*Psi';
sat = @(x) sign(x).*min(abs(x),1);
EPS = 1e-5;
MAX_ITER = 100000;
z = [zeros(2*n,1);zeta]; v = zeros(mu,1);
r = zeta;
k = 0;
gamma = 0.05;
while (norm(r)>EPS) & (k < MAX_ITER)
    u = M*(z-v);
    z0 = soft_thresholding(gamma,u+v(1:n));
    z1 = sat(u+v(n+1:2*n));
    z2 = zeta;
    z = [z0;z1;z2];
    v = v + Psi*u - z;
    r = Phi*u - zeta;
    k = k + 1;
end
%% Plot
figure;
plot(0:T/n:T-T/n,u,'LineWidth',2);
title('Sparse control');
```

9.6 Further Reading

The time discretization discussed in this section is based on the fundamental theory of sampled-data control, for which you can refer to a standard textbook by Chen and Francis [22]. The concept of pathological sampling is also found in this book.

Instead of the time discretization method, one can also use the *shooting method* for numerical computation of L^1-optimal control. The shooting method is based on the necessary conditions by Pontryagin's minimum principle. For the shooting method, see [10] for details.

DOI: 10.1561/9781680837254.ch10

Chapter 10

Advanced Topics

In this chapter, we introduce advanced topics in maximum hands-off control.

10.1 Smooth Hands-off Control by Mixed L^1/L^2 Optimization

As we studied in Chapter 8, the maximum hands-off control (the L^0-optimal control) is bang-off-bang (Theorem 8.14, p. 162), that is, it is a piecewise constant function taking values of ± 1 or 0. This means that the maximum hands-off control is *discontinuous*; the control changes its value between 1 and 0, or 0 and -1 at switching times. This is undesirable for some applications in which the actuators cannot move abruptly. In this case, one may want to make the control *continuous*. For this purpose, we add a regularization term to the L^1 cost $J_1(u)$ in the L^1 optimal control problem (L^1-OPT) (p. 155). That is, we consider the following cost function:

$$J_{12}(u) = \lambda \|u\|_1 + \frac{1}{2}\|u\|_2^2 = \int_0^T \left(\lambda|u(t)| + \frac{1}{2}|u(t)|^2\right)dt, \qquad (10.1)$$

where $\lambda > 0$ is a fixed parameter.

The idea to add the L^2-norm term is borrowed by the *elastic net regularization*[1] in compressed sensing [119]. The elastic net regularization promotes sparsity with the grouping effect, where strongly correlated vectors are chosen at the same time. This ensures that the solution is not overly sensitive to small changes in the observation. From this idea, the L^2 term in (10.1) enhances continuity of the solution.

With the cost function (10.1), we consider the following mixed L^1/L^2-optimal control problem.

L^1/L^2-optimal control problem (L^1/L^2-OPT)

For the linear time-invariant system

$$\dot{x}(t) = Ax(t) + bu(t), \quad t \geq 0, \quad x(0) = \xi \in \mathbb{R}^d,$$

find a control $\{u(t) : t \in [0, T]\}$ with $T > 0$ that minimizes

$$J_{12}(u) = \lambda \|u\|_1 + \frac{1}{2}\|u\|_2^2$$

subject to

$$x(T) = 0,$$

and

$$\|u\|_\infty \leq 1.$$

To discuss properties of the L^1/L^2-optimal control, we give necessary conditions of optimality by Pontryagin's minimum principle.

The Hamiltonian function associated to (L^1/L^2-OPT) is given by

$$H^\eta(x, p, u) = p^\top(Ax + bu) + \eta\left(\lambda|u| + \frac{1}{2}|u|^2\right). \tag{10.2}$$

We do not consider the abnormal case (i.e., $\eta = 0$) and assume $\eta = 1$. Let $u^*(t)$ denote the optimal control and $x^*(t)$ and $p^*(t)$ the resultant optimal state and costate, respectively. Then we have the following result.

Lemma 10.1. *The L^1/L^2-optimal control $u^*(t)$ satisfies*

$$u^*(t) = -\text{sat}\left(S_\lambda\left(p^*(t)^\top b\right)\right), \tag{10.3}$$

---

1.   The name "elastic net" is meant to suggest a stretchable fishing net that retains all the big fish.

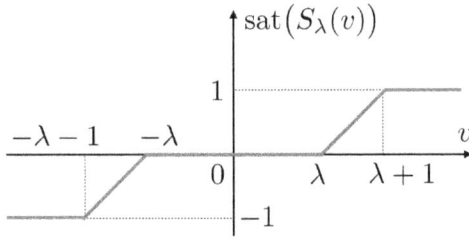

Figure 10.1. Saturated shrinkage function $\mathrm{sat}(S_\lambda(v))$.

where $S_\lambda(\cdot)$ is the soft-thresholding operator (see Section 4.2.5, p. 69) defined by

$$
S_\lambda(v) \triangleq
\begin{cases}
v + \lambda & \text{if } v < -\lambda, \\
0, & \text{if } -\lambda \le v \le \lambda, \\
v - \lambda, & \text{if } \lambda < v,
\end{cases}
\tag{10.4}
$$

and $\mathrm{sat}(\cdot)$ *is the saturation function defined by*

$$
\mathrm{sat}(v) \triangleq
\begin{cases}
-1, & \text{if } v < -1, \\
v, & \text{if } -1 \le v \le 1, \\
1, & \text{if } 1 < v.
\end{cases}
\tag{10.5}
$$

See Figure 10.1 for the graphs of $\mathrm{sat}(S_\lambda(v))$ *in (10.3).*

Proof: From Pontryagin's minimum principle, we have

$$
u^*(t) = \underset{u \in [-1,1]}{\arg\min} \left\{ \left(p^*(t)^\top b\right)u + \lambda|u| + \frac{1}{2}|u|^2 \right\}
$$

$$
=
\begin{cases}
1, & \text{if } p^*(t)^\top b \le -\lambda - 1 \\
-\left(p^*(t)^\top b + \lambda\right), & \text{if } -\lambda - 1 < p^*(t)^\top b < -\lambda \\
0, & \text{if } -\lambda \le p^*(t)^\top b \le \lambda \\
-\left(p^*(t)^\top b - \lambda\right), & \text{if } \lambda < p^*(t)^\top b < \lambda + 1 \\
-1, & \text{if } \lambda + 1 \le p^*(t)^\top b
\end{cases}
\tag{10.6}
$$

$$
= -\mathrm{sat}\left(S_\lambda\left(p^*(t)^\top b\right) \right).
$$

$\square$

From Lemma 10.1, we have the following theorem.

Theorem 10.2 (Continuity). *The L^1/L^2-optimal control $u^*(t)$ is continuous in t over $[0, T]$.*

Proof: Define

$$\bar{u}(p) \triangleq -\mathrm{sat}\left(S_\lambda\left(p^\top b\right)\right). \tag{10.7}$$

Since the composite function sat $\circ$ S_λ is continuous (see Figure 10.1), $\bar{u}(p)$ is also continuous in p. It follows from Lemma 10.1 that the optimal control u^* given in (10.3) is continuous in p^*. Hence, $u^*(t)$ is continuous, if $p^*(t)$ is continuous in t over $[0, T]$. In fact, from (8.20) (p. 158), $p^*(t)^\top b$ is given by

$$p^*(t)^\top b = p^*(0)^\top e^{-At} b, \tag{10.8}$$

which is continuous in t over $\mathbb{R}$.   □

Theorem 10.2 motivates us to use the L^1/L^2 optimization in the L^1/L^2-optimal control problem (L^1/L^2-OPT) for continuous hands-off control.

In general, the degree of continuity (or smoothness) and the sparsity of the control input cannot be optimized at the same time. The weight parameter λ can be used for trading smoothness for sparsity. Lemma 10.1 suggests that increasing the weight λ makes the L^1/L^2 optimal control $u^*(t)$ sparser (see also Figure 10.1). On the other hand, decreasing λ smoothens $u^*(t)$.

Example 10.3. *Let us consider the following linear system*

$$\frac{d\boldsymbol{x}(t)}{dt} = \begin{bmatrix} 0 & 1 & 0 & 0 \\ 0 & 0 & 1 & 0 \\ 0 & 0 & 0 & 1 \\ 0 & 0 & 0 & 0 \end{bmatrix} \boldsymbol{x}(t) + \begin{bmatrix} 0 \\ 0 \\ 0 \\ 1 \end{bmatrix} u(t). \tag{10.9}$$

We set the final time $T = 10$, and the initial and final states as

$$\boldsymbol{x}(0) = [0.5, 0.5, 0.5, 0.5]^\top, \quad \boldsymbol{x}(10) = \boldsymbol{0}. \tag{10.10}$$

Figure 10.2 shows the L^1/L^2 optimal control with weights $\lambda = 1$. The maximum hands-off control is also illustrated. We can see that the L^1/L^2-optimal control is continuous but sufficiently sparse.

10.2   Discrete-valued Control

As we observed in Chapter 8, the maximum hands-off control (or the L^0-optimal control) takes values in an *alphabet*[2] $\{-1, 0, 1\}$. Such a control is called a *discrete-valued control*, since the control takes a finite number of values. Discrete-valued

---

2.   The word *alphabet* is borrowed from information theory [27]. An alphabet is a set of a finite number of elements that are used to represent signals of interest.

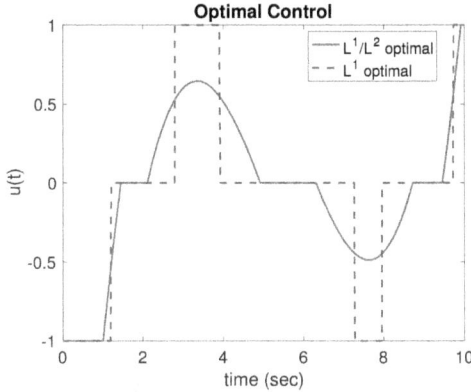

Figure 10.2. Maximum hands-off control (dashed) and L^1/L^2-optimal control (solid).

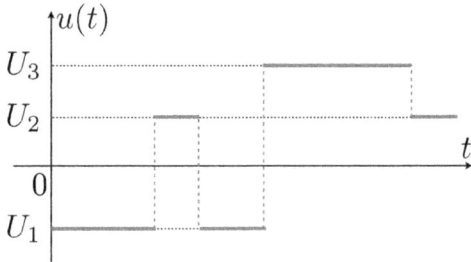

Figure 10.3. An example of discrete-valued control that takes three values of U_1, U_2, and U_3.

control is important in networked control systems where the bandwidth of the network is limited, since discrete-valued signals can be effectively compressed.

We here generalize the property of discreteness in maximum hands-off control by the sum-of-absolute-values (SOAV) optimization.

10.2.1 Sum-of-Absolute-Values (SOAV) Optimization

Let us consider discrete-valued control for the linear time-invariant plant

$$\dot{x}(t) = Ax(t) + bu(t), \quad t \geq 0, \quad x(0) = \xi \in \mathbb{R}^d, \tag{10.11}$$

where the control $u(t)$ takes N real numbers

$$U_1 < U_2 < \cdots < U_N. \tag{10.12}$$

That is, we consider a discrete-valued control with alphabet $\{U_1, U_2, \ldots, U_N\}$. Figure 10.3 shows an example of discrete-valued control. We then seek a discrete-valued control that achieves $x(T) = 0$, given the initial state $x(0) = \xi$ and the control time $T > 0$.

A standard method to obtain discrete-valued control is to describe the problem as a mixed-integer programming problem [6]. However, this method requires a lot of computational time, which glows exponentially as the size of problem glows, and hence this method is hard to apply for a large scale problem. Instead, we consider convex relaxation of this optimization problem of discrete-valued control.

We first define the *feasible controls* that drive the state $x(t)$ from the initial state $x(0) = \xi$ to the origin in time $T > 0$, satisfying

$$U_1 \leq u(t) \leq U_N, \quad \forall t \in [0, T]. \tag{10.13}$$

We denote by $\mathcal{U}(T, \xi)$ the set of feasible controls. We assume that $\xi \in \mathbb{R}^d$ and $T > 0$ are given such that $\mathcal{U}(T, \xi)$ is non-empty. For a feasible control $u \in \mathcal{U}(T, \xi)$, define the following cost function:

$$J_0(u) \triangleq \sum_{j=1}^{N} w_j \|u - U_j\|_0, \tag{10.14}$$

where $w_1, w_2, \ldots, w_N$ are weights that satisfy

$$w_i > 0, \quad w_1 + w_2 + \cdots + w_N = 1. \tag{10.15}$$

Minimizing the cost function (10.14) may promote discreteness of the control to take values in $\{U_1, \ldots, U_N\}$. This can be explained as follows. A discrete-valued control is a piecewise constant signal as shown in Figure 10.3. If $u(t) = U_j$ for t in some time intervals with a positive length, then the function $u(t) - U_j$ is zero over the intervals, and hence it is *sparse*. Namely, the L^0 norm of the function $u - U_j$ should be smaller than T. If we choose the weights $w_1, \ldots, w_N$ according to the importance of the values $U_1, \ldots, U_N$ and minimize the cost function (10.14), we may obtain a discrete-valued feasible control.

The cost function (10.14) is discontinuous and non-convex, and hence it is difficult to directly obtain the optimal solution as in the case of L^0-optimal control. We then adopt the L^1 relaxation, that is, we use the L^1 norm instead of the L^0 norm in (10.14):

$$J_1(u) \triangleq \sum_{j=1}^{N} w_j \|u - U_j\|_1 = \int_0^T \sum_{j=1}^{N} w_j |u(t) - U_j| dt \tag{10.16}$$

We call this cost function the *sum of absolute values* or *SOAV* for short, and the optimal control that minimizes the SOAV cost function the *sum-of-absolute-values optimal control* or *SOAV-optimal control*. We describe the SOAV-optimal control problem as follows:

─────────── SOAV-optimal control problem (SOAV-OPT) ───────────

For the linear time-invariant system

$$\dot{x}(t) = Ax(t) + bu(t), \quad t \geq 0, \quad x(0) = \xi \in \mathbb{R}^d,$$

find a control $\{u(t) : t \in [0, T]\}$ that minimizes

$$J_1(u) = \sum_{j=1}^{N} w_j \|u - U_j\|_1 = \int_0^T \sum_{j=1}^{N} w_j |u(t) - U_j| dt$$

subject to

$$x(T) = 0,$$

and

$$U_1 \leq u(t) \leq U_N, \quad \forall t \in [0, T].$$

10.2.2   Discreteness of SOAV-optimal Control

Here we show that the SOAV-optimal control is a discrete-valued control taking values in $\{U_1, \ldots, U_N\}$ under some conditions.

Let $u^* \in \mathcal{U}(T, \xi)$ be an SOAV-optimal control minimizing the cost function (10.16), that is,

$$u^* = \arg\min_u \ J_1(u) \text{ subject to } u \in \mathcal{U}(T, \xi). \tag{10.17}$$

For the optimal control problem (SOAV-OPT), we analyze the solution u^* by using Pontryagin's minimum principle.

The stage-cost function $L(u)$ of the SOAV cost function (10.16) is given by

$$L(u) = \sum_{j=1}^{N} w_j |u - U_j|. \tag{10.18}$$

Figure 10.4 shows an example of function $L(u)$. As shown in this figure, the stage-cost function $L(u)$ is a continuous and piecewise linear function. Also, since the function $L(u)$ is a convex combination of convex functions $|u - U_j|$, $j = 1, \ldots, N$, $L(u)$ is convex in u. That is, the optimization problem in (SOAV-OPT) is a convex optimization problem.

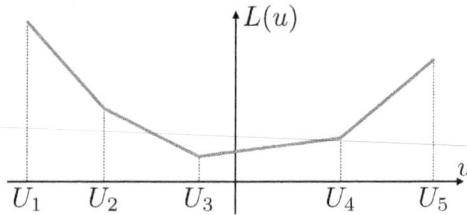

Figure 10.4. Piecewise linear function $L(u)$.

Then the Hamiltonian for (SOAV-OPT) is defined by

$$H^{\eta}(\boldsymbol{x}, \boldsymbol{p}, u) = \boldsymbol{p}^{\top}(A\boldsymbol{x} + \boldsymbol{b}u) + \eta L(u)$$

$$= \boldsymbol{p}^{\top}(A\boldsymbol{x} + \boldsymbol{b}u) + \eta \sum_{j=1}^{N} w_j |u - U_j|. \tag{10.19}$$

Here we assume[3] $\eta = 1$. Let $\boldsymbol{x}^*$ and $\boldsymbol{p}^*$ be respectively the optimal state and costate with the optimal control u^*. From the minimum principle, we have

$$u^*(t) = \underset{u \in [U_1, U_N]}{\arg\min} \left\{ \boldsymbol{p}^*(t)^{\top}(A\boldsymbol{x}^*(t) + \boldsymbol{b}u) + L(u) \right\}$$

$$= \underset{u \in [U_1, U_N]}{\arg\min} \left\{ \boldsymbol{p}^*(t)^{\top} \boldsymbol{b}u + L(u) \right\}. \tag{10.20}$$

Let us solve the minimization problem in (10.20).

Since the function $L(u)$ is piecewise linear, $L(u)$ can be written as

$$L(u) = \begin{cases} a_1 u + b_1, & u \in [U_1, U_2], \\ a_2 u + b_2, & u \in [U_2, U_3], \\ \quad \vdots \\ a_{N-1} u + b_{N-1}, & u \in [U_{N-1}, U_N], \end{cases} \tag{10.21}$$

where

$$a_k = \sum_{j=1}^{k} w_j - \sum_{j=k+1}^{N} w_j,$$

$$b_k = -\sum_{j=1}^{k} w_j U_j + \sum_{j=k+1}^{N} w_j U_j, \quad k = 1, 2, \ldots, N-1. \tag{10.22}$$

---

3.   If $\eta = 0$, then the extremum solution is a bang-bang control that takes values of U_1 or U_N, which is a discrete-valued control.

Fix $t \in [0, T]$ and define $\alpha \triangleq p^*(t)^\top b \in \mathbb{R}$. Since $L(u)$ is continuous, and the following inequality

$$a_1 < a_2 < \cdots < a_{N-1} \tag{10.23}$$

holds, we can compute the minimizer of

$$h(u) \triangleq \alpha u + L(u) = \begin{cases} (a_1 + \alpha)u + b_1, & u \in [U_1, U_2], \\ (a_2 + \alpha)u + b_2, & u \in [U_2, U_3], \\ \vdots \\ (a_{N-1} + \alpha)u + b_{N-1}, & u \in [U_{N-1}, U_N], \end{cases} \tag{10.24}$$

for $u \in [U_1, U_N]$.

(i) If $a_1 + \alpha > 0$, then from (10.23) we have

$$0 < a_1 + \alpha < a_2 + \alpha < \cdots < a_{N-1} + \alpha, \tag{10.25}$$

and the slopes $(a_k + \alpha)$ of the linear functions in (10.24) are all positive. See (i) of Figure 10.5. Hence we have

$$\arg\min_{u \in [U_1, U_N]} h(u) = U_1. \tag{10.26}$$

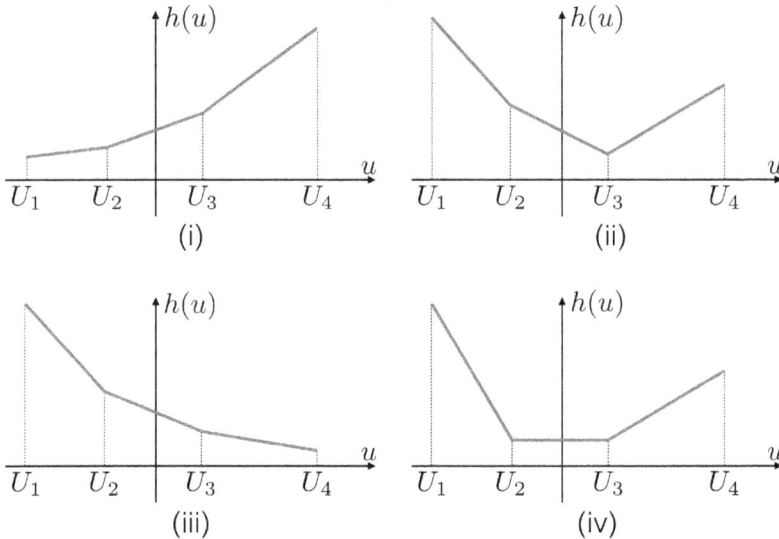

Figure 10.5. 4 cases of piecewise linear function $h(u) = \alpha u + L(u)$.

(ii) If $a_k + \alpha < 0$ and $a_{k+1} + \alpha > 0$ ($k = 1, \ldots, N - 2$), then from (10.23) we have

$$a_1 + \alpha < a_2 + \alpha < \cdots < a_k + \alpha < 0, \qquad (10.27)$$

and

$$0 < a_{k+1} + \alpha < a_{k+2} + \alpha < \cdots < a_{N-1} + \alpha. \qquad (10.28)$$

The sign of the slopes of the linear functions in (10.24) changes from negative to positive at $u = U_{k+1}$. See (ii) of Figure 10.5. Hence, we have

$$\arg\min_{u \in [U_1, U_N]} h(u) = U_{k+1}. \qquad (10.29)$$

(iii) If $a_{N-1} + \alpha < 0$, then we have

$$a_1 + \alpha < a_2 + \alpha < \cdots < a_{N-1} + \alpha < 0, \qquad (10.30)$$

and the slopes $(a_k + \alpha)$ in (10.24) are all negative. See (iii) of Figure 10.5. Hence we have

$$\arg\min_{u \in [U_1, U_N]} h(u) = U_N. \qquad (10.31)$$

(iv) If there exists $k \in \{1, 2, \ldots, N - 1\}$ such that $a_k + \alpha = 0$, then the slope becomes zero over the interval $[U_k, U_{k+1}]$. Hence we have

$$\arg\min_{u \in [U_1, U_N]} h(u) = [U_k, U_{k+1}]. \qquad (10.32)$$

In this case, we cannot determine the unique value for $u^*(t)$.

In summary, the SOAV optimal control $u^*(t)$ satisfies the following:

$$u^*(t) = \begin{cases} U_1, & \text{if } -a_1 < p^*(t)^\top b \\ U_2, & \text{if } -a_2 < p^*(t)^\top b < -a_1 \\ \quad\vdots \\ U_{N-1}, & \text{if } -a_{N-1} < p^*(t)^\top b < -a_{N-2} \\ U_N, & \text{if } p^*(t)^\top b < -a_{N-1} \end{cases} \qquad (10.33)$$

$$u^*(t) \in [U_k, U_{k+1}], \quad \text{if } p^*(t)^\top b = -a_k, \quad k = 1, 2, \ldots, N - 1 \qquad (10.34)$$

From (10.34), if

$$p^*(t)^\top b \neq -a_k, \quad k = 1, 2, \ldots, N - 1 \qquad (10.35)$$

holds for almost all $t \in [0, T]$, then we can see that $u^*(t)$ takes discrete-values in $\{U_1, \ldots, U_N\}$ for almost all $t \in [0, T]$. Let us consider a sufficient condition for this.

We see that (10.35) holds for almost all $t \in [0, T]$ if and only if

$$\mu\left(\{t \in [0, T] : p^*(t)^\top b = -a_k\}\right) = 0 \qquad (10.36)$$

for $k = 1, 2, \ldots, N - 1$. We say the SOAV optimal control is *non-singular* if (10.36) holds. Then we have the following theorem:

Theorem 10.4. *Assume that the SOAV optimal control is non-singular. Then the optimal control $u^*(t)$ takes values in $\{U_1, \ldots, U_N\}$ for almost all $t \in [0, T]$.*

For the non-singularity, we have the following theorem.

Theorem 10.5. *Assume that the pair (A, b) is non-singular.[4] Assume also that*

$$\sum_{j=1}^{k} w_j \neq \sum_{j=k+1}^{N} w_j \qquad (10.37)$$

holds for $k = 1, 2, \ldots, N - 1$. Then the SOAV optimal control is non-singular.

Exercise 10.6. Prove Theorem 10.5.

The condition (10.37) in Theorem 10.5 is a sufficient and necessary condition for the slopes of the linear functions in (10.24) to be nonzero.

Example 10.7. *Let us consider a design example of SOAV-optimal control. We consider the 4-th order plant given in (10.9) in Example 10.3. The final time $T = 10$ and the initial and final states are the same as (10.10).*
The alphabet is given by $\{-1, -0.5, 0, 0.5, 1\}$, that is, $N = 5$ and

$$U_1 = -1, \; U_2 = -0.5, \; U_3 = 0, \; U_4 = 0.5, \; U_5 = 1. \qquad (10.38)$$

The weights in the cost function (10.16) are set as

$$w_1 = w_2 = w_3 = w_4 = w_5 = \frac{1}{5}. \qquad (10.39)$$

Figure 10.6 shows the obtained SOAV-optimal control. In this figure, the maximum hands-off control (L^1-optimal control) discussed in Chapter 8 and the

---

4.   The pair (A, b) is said to be non-singular if the pair (A, b) is controllable and A is non-singular.

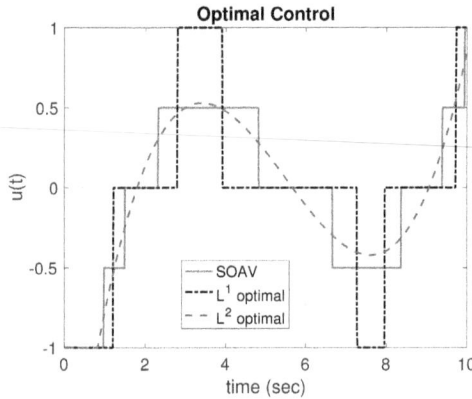

Figure 10.6. SOAV-optimal control (solid), L^1-optimal control (dashed), and L^2-optimal control (dotted).

L^2-optimal control[5] that minimizes the L^2 norm

$$J_2(u) = \int_0^T |u(t)|^2 dt \tag{10.40}$$

are also shown. Note that the L^1-optimal control is bang-off-bang and takes values of ± 1 and 0. On the other hand, the L^2-optimal control is a smooth control. The SOAV-optimal control is between them. It takes discrete values in the alphabet $\{-1, -0.5, 0, 0.5, 1\}$, that is a quantization of the L^2-optimal control.

Figure 10.7 shows the state variables $x_1(t), \ldots, x_4(t)$ in the state $x(t)$ and the SOAV-optimal control. We can see that by the obtained discrete-valued control $u(t)$, all the state variables converge to the origin in the time $T = 10$. Note that this cannot be possible when one uses a quantized version of the L^2-optimal control by a static quantizer; there should be quantization errors that perturb the state trajectory.

10.3   Time-Optimal Hands-off Control

In this section, we consider an optimal control that takes account of sparsity and time-optimality at the same time. Let us consider the following linear time-invariant system:

$$\dot{x}(t) = Ax(t) + bu(t), \quad t \geq 0, \quad x(0) = \xi \in \mathbb{R}^d. \tag{10.41}$$

---

5.   The L^2-optimal control is also known as *minimum energy control* [2, Section 6-18].

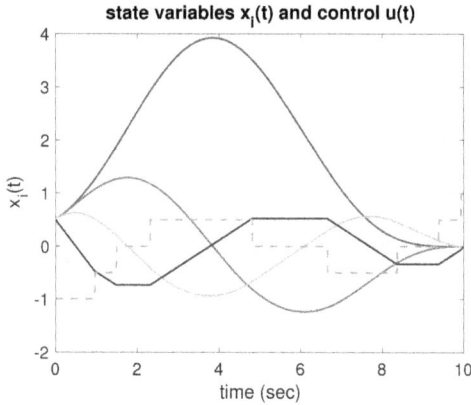

Figure 10.7. State variables $x_1(t), \ldots, x_4(t)$ and the SOAV-optimal control $u(t)$, $t \in [0, 10]$.

The control objective is to drive the state to the origin. Here we do not fix the final time T. As in the minimum-time control in Chapter 7, the final time T is also an optimization variable.

First, we consider the feasibility of the control. For the system (10.41), a control u is said to be feasible if there exists a finite time $T > 0$ such that by $\{u(t) : t \in [0, T]\}$ satisfying

$$|u(t)| \leq 1, \quad \forall t \in [0, T], \tag{10.42}$$

the state $x(t)$ in (10.41) is steered from $x(0) = \xi$ to $x(T) = 0$. From the definition of the controllable set $\mathcal{R}$ in (7.30) (p. 139), there exists a feasible control if the initial state ξ is in the controllable set $\mathcal{R}$. Therefore, we assume $\xi \in \mathcal{R}$. Using the feasible set $\mathcal{U}(T, \xi)$ with fixed $T > 0$ (see Section 7.1.3), the set of all feasible controls is given by

$$\mathcal{U}(\xi) = \bigcup_{T \geq 0} \mathcal{U}(T, \xi). \tag{10.43}$$

Next, we formulate the optimal control problem. We seek a feasible control $u \in \mathcal{U}(\xi)$ that minimizes the L^0 norm of u and the response time T at the same time. For this, we consider the following cost function:

$$J_0(u) \triangleq \lambda \|u\|_0 + T, \tag{10.44}$$

where $\lambda > 0$ is a weight parameter for a trade-off between the two requirements. As usual, we relax the L^0 norm in (10.44) by the L^1 norm $\|u\|_1$, namely, we consider the following cost function:

$$J_1(u) \triangleq \lambda \|u\|_1 + T, \tag{10.45}$$

Now we formulate our problem.

L^1-time-optimal control problem (L^1-T-OPT)

For the linear time-invariant system

$$\dot{x}(t) = Ax(t) + bu(t), \quad t \geq 0, \quad x(0) = \xi \in \mathbb{R}^d,$$

find a control $\{u(t) : t \in [0, \infty)\}$ that minimizes

$$J_1(u) = \lambda\|u\|_1 + T$$

subject to

$$x(T) = 0,$$

and

$$\|u\|_\infty \leq 1.$$

We call the optimal solution the *L^1-time-optimal control*.

The existence theorem for the L^1-time-optimal control is proved similarly to the time-optimal control (Theorem 8.15, p. 163).

Theorem 10.8. *For any initial state $\xi \in \mathcal{R}$, there exists at least one L^1-time-optimal control.*

You can find the proof in [57].

The Hamiltonian for (L^1-T-OPT) is given by

$$H^\eta(x, p, u) = p^\top(Ax + bu) + \eta(\lambda|u| + 1) \tag{10.46}$$

We do not consider the abnormal case ($\eta = 0$) and assume $\eta = 1$. Then the optimal control $u^*(t)$ of (L^1-T-OPT) satisfies

$$u^*(t) = \arg\min_{u\in[-1,1]} H^1(x, p, u) = \arg\min_{u\in[-1,1]} \left\{ p^*(t)^\top bu + \lambda|u| \right\} \tag{10.47}$$

From this, we have

$$u^*(t) = \begin{cases} 1, & \text{if } p^*(t)^\top b < -\lambda, \\ 0, & \text{if } -\lambda < p^*(t)^\top b < \lambda, \\ -1, & \text{if } \lambda < p^*(t)^\top b, \end{cases} \tag{10.48}$$

$$u^*(t) \in [0, 1], \quad \text{if } p^*(t)^\top b = -\lambda,$$
$$u^*(t) \in [-1, 0], \quad \text{if } p^*(t)^\top b = \lambda.$$

If $p^*(t)^\top b = \pm\lambda$ holds only on sets of measure zero (i.e., if $p^*(t)^\top b \neq \pm\lambda$ almost all $t \in [0, T]$, then the L^1-time-optimal control is bang-off-bang. From Lemma 8.3, we have the following theorem:

Theorem 10.9. *Assume that the pair* (A, b) *is non-singular. Then the* L^1*-time-optimal control is bang-off-bang (if it exists).*

This theorem along with Theorem 10.8 leads to the following equivalence theorem:

Theorem 10.10. *Assume* $\xi \in \mathcal{R}$ *and the pair* (A, b) *is non-singular. Then the* L^1*-time-optimal control is equivalent to the* L^0*-time-optimal control that minimizes the cost function (10.44).*

10.4   Further Reading

The smooth hands-off control by the mixed L^1/L^2 optimization was first proposed in [83]. Another formulation for smooth hands-off control by the CLOT (Combined L-One and Two) norm was also proposed in [85]. The CLOT norm is defined by

$$\|u\|_{\mathrm{CLOT}} \triangleq \lambda_1 \|u\|_1 + \lambda_2 \|u\|_2, \tag{10.49}$$

with parameters $\lambda_1 > 0$ and $\lambda_2 > 0$ such that $\lambda_1 + \lambda_2 = 1$. Compared with the mixed L^1/L^2 cost function in (10.1), the L^2 term in the CLOT norm is not squared. The CLOT-optimal control is also continuous but sparser than the mixed L^1/L^2-optimal control in (L^1/L^2-OPT).

The SOAV optimal control has been proposed in [53, 56]. The idea of SOAV cost function was first proposed in [77] for discrete-valued signal reconstruction. The SOAV optimization was then applied to digital communications [46, 101, 102].

The L^1-time-optimal control was first proposed in [57].

References

[1] M. Aldridge, L. Baldassini, and O. Johnson, "Group testing algorithms: Bounds and simulations," *IEEE Trans. Inf. Theory*, vol. 60, no. 6, pp. 3671–3687, Jun. 2014.

[2] M. Athans and P. L. Falb, *Optimal Control*. Dover Publications, 2007, an unabridged republication of the work published by McGraw-Hill in 1966.

[3] G. K. Atia and V. Saligrama, "Boolean compressed sensing and noisy group testing," *IEEE Trans. Inf. Theory*, vol. 58, no. 3, pp. 1880–1901, Mar. 2012.

[4] H. H. Bauschke and P. L. Combettes, *Convex Analysis and Monotone Operator Theory in Hilbert Spaces*. Springer, 2011.

[5] A. Beck and M. Teboulle, "Gradient-based algorithms with applications to signal-recovery problems," in *Convex Optimization*. Cambridge University Press, 2010.

[6] A. Bemporad and M. Morari, "Control of systems integrating logic, dynamics, and constraints," *Automatica*, vol. 35, pp. 407–427, 1999.

[7] D. Bertsekas, *Convex Optimization Algorithms*. Athena Scientific, 2015.

[8] C. M. Bishop, *Pattern Recognition and Machine Learning*. Springer, 2006.

[9] T. Blumensath and M. E. Davies, "Iterative thresholding for sparse approximations," *Journal of Fourier Analysis and Applications*, vol. 14, no. 5, pp. 629–654, 2008.

[10] H. Bock and K. Plitt, "A multiple shooting algorithm for direct solution of optimal control problems*," *IFAC Proceedings Volumes*, vol. 17, no. 2, pp. 1603–1608, 1984.

[11] S. Boyd, L. E. Ghaoui, e. Feron, and V. Balakrishnan, *Linear Matrix Inequalities in System and Control Theory*. SIAM, 1994.

[12] S. Boyd, N. Parikh, E. Chu, B. Peleato, and J. Eckstein, "Distributed optimization and statistical learning via the alternating direction method of multipliers," *Foundations and Trends in Machine Learning*, vol. 3, no. 1, pp. 1–122, 2011.

[13] S. Boyd and L. Vandenberghe, *Convex Optimization*. Cambridge University Press, 2004.

[14] J. P. Boyle and R. L. Dykstra, "A method for finding projections onto the intersection of convex sets in Hilbert spaces," in *Advances in Order Restricted Statistical Inference, Lecture Notes in Statistics*, R. Dykstra, T. Robertson, and F. T. Wright, Eds. New York: Springer, 1986, vol. 37.

[15] P. Bühlmann and S. van de Geer, *Statistics for High-Dimensional Data*. Springer, 2011.

[16] E. J. Candes and T. Tao, "Near-optimal signal recovery from random projections: Universal encoding strategies?" *IEEE Trans. Inf. Theory*, vol. 52, no. 12, pp. 5406–5425, Dec. 2006.

[17] C. Chan, "The state of the art of electric, hybrid, and fuel cell vehicles," *Proc. IEEE*, vol. 95, no. 4, pp. 704–718, Apr. 2007.

[18] C. Chang and S. Sim, "Optimising train movements through coast control using genetic algorithms," *IEE Proceedings-Electric Power Applications*, vol. 144, no. 1, pp. 65–73, 1997.

[19] D. Chatterjee, M. Nagahara, D. E. Quevedo, and K. M. Rao, "Characterization of maximum hands-off control," *Systems & Control Letters*, vol. 94, pp. 31–36, 2016.

[20] S. S. Chen, D. L. Donoho, and M. A. Saunders, "Atomic decomposition by basis pursuit," *SIAM J. Sci. Comput.*, vol. 20, no. 1, pp. 33–61, Aug. 1998.

[21] S. Chen and D. Donoho, "Basis pursuit," in *Signals, Systems and Computers, Conference Record of the Twenty-Eighth Asilomar Conference on*, vol. 1, Oct. 1994, pp. 41–44.

[22] T. Chen and B. A. Francis, *Optimal Sampled-Data Control Systems*. Springer, 1995.

[23] J. F. Claerbout and F. Muir, "Robust modeling with erratic data," *Geophysics*, vol. 38, no. 5, pp. 826–844, 1973.

[24] F. Clarke, *Functional Analysis, Calculus of Variations and Optimal Control*, ser. Graduate Texts in Mathematics. Springer, London, 2013, vol. 264.

[25] P. L. Combettes and J.-C. Pesquet, "Proximal splitting methods in signal processing," in *Fixed-Point Algorithms for Inverse Problems in Science and Engineering*. New York, NY: Springer New York, 2011, pp. 185–212.

[26] T. H. Cormen, C. E. Leiserson, R. L. Rivest, and C. Stein, *Introduction to Algorithms*, 3rd ed. MIT Press, 2009.

[27] T. M. Cover and J. A. Thomas, *Elements of Information Theory*, 2nd ed. Wiley–Interscience, 2006.

[28] G. M. Davis, S. G. Mallat, and Z. Zhang, "Adaptive time-frequency decompositions," *Optical Engineering*, vol. 33, no. 7, pp. 2183–2191, 1994.

[29] N. K. Dhingra, M. R. Jovanović, and Z. Luo, "An ADMM algorithm for optimal sensor and actuator selection," in *53rd IEEE Conference on Decision and Control*, 2014, pp. 4039–4044.

[30] D. L. Donoho, "Compressed sensing," *IEEE Trans. Inf. Theory*, vol. 52, no. 4, pp. 1289–1306, Apr. 2006.

[31] D. L. Donoho and P. B. Stark, "Uncertainty principles and signal recovery," *SIAM Journal on Applied Mathematics*, vol. 49, no. 3, pp. 906–931, 1989.

[32] R. Dorfman, "The detection of defective members of large populations," *Ann. Math. Statist.*, vol. 14, no. 4, pp. 436–440, 12 1943.

[33] G.-R. Duan and H.-H. Yu, *LMIs in Control Systems*. CRC Press, 2013.

[34] B. Dunham, "Automatic on/off switching gives 10-percent gas saving," *Popular Science*, vol. 205, no. 4, p. 170, Oct. 1974.

[35] J. Eckstein and D. Bertsekas, "On the Douglas-Rachford splitting method and proximal point algorithm for maximal monotone operators," *Math. Program.*, vol. 55, pp. 293–318, 1992.

[36] M. B. Egerstedt and C. F. Martin, *Control Theoretic Splines: Optimal Control, Statistics, and Path Planning*. Princeton University Press, 2009.

[37] M. Elad, *Sparse and Redundant Representations*. Springer, 2010.

[38] S. Foucart and H. Rauhut, *A Mathematical Introduction to Compressive Sensing*. Birkhäuser, 2013.

[39] M. Gallieri and J. M. Maciejowski, "ℓ_{asso} MPC: Smart regulation of over-actuated systems," in *Proc. Amer. Contr. Conf.*, Jun. 2012, pp. 1217–1222.

[40] C. Giraud, *Introduction to High-Dimensional Statistics*. CRC Press, 2015.

[41] G. H. Golub and C. F. V. Loan, *Matrix Computations*, 4th ed. Johns Hopkins University Press, 2012.

[42] I. Goodfellow, Y. Bengio, and A. Courville, *Deep Learning*. MIT Press, 2016.

[43] D. A. Harville, *Matrix Algebra From a Statistician's Perspective*. Springer, 1997.

[44] T. Hastie, R. Tibshirani, and J. Friedman, *The Elements of Statistical Learning*. Springer, 2009.

[45] T. Hastie, R. Tibshirani, and M. Wainwright, *Statistical Learning with Sparsity: The Lasso and Generalizations*. CRC Press, 2015.

[46] R. Hayakawa and K. Hayashi, "Discreteness-aware approximate message passing for discrete-valued vector reconstruction," *IEEE Trans. Signal Process.*, vol. 66, no. 24, pp. 6443–6457, 2018.

[47] K. Hayashi, M. Nagahara, and T. Tanaka, "A user's guide to compressed sensing for communications systems," *IEICE Trans. on Communications*, vol. E96-B, no. 3, pp. 685–712, Mar. 2013.

[48] W. P. M. H. Heemels, K. H. Johansson, and P. Tabuada, "An introduction to event-triggered and self-triggered control," in *2012 IEEE 51st IEEE Conference on Decision and Control (CDC)*, Dec. 2012, pp. 3270–3285.

[49] H. Hermes and J. P. Lasalle, *Functional Analysis and Time Optimal Control*. Academic Press, 1969.

[50] T. Ikeda and K. Kashima, "Sparsity-constrained controllability maximization with application to time-varying control node selection," *IEEE Control Systems Letters*, vol. 2, pp. 321–326, 2018.

[51] T. Ikeda and K. Kashima, "On sparse optimal control for general linear systems," *IEEE Trans. Autom. Control*, vol. 64, no. 5, pp. 2077–2083, 2019.

[52] T. Ikeda, M. Nagahara, and K. Kashima, "Maximum hands-off distributed control for consensus of multi-agent systems with sampled-data state observation," *IEEE Trans. Control Netw. Syst.*, vol. 6, no. 2, pp. 852–862, Jun. 2019.

[53] T. Ikeda, M. Nagahara, and S. Ono, "Discrete-valued control of linear time-invariant systems by sum-of-absolute-values optimization," *IEEE Trans. Autom. Control*, vol. 62, no. 6, pp. 2750–2763, 2017.

[54] T. Ikeda, D. Zelazo, and K. Kashima, "Maximum hands-off distributed bearing-based formation control," in *2019 IEEE 58th Conference on Decision and Control (CDC)*, 2019, pp. 4459–4464.

[55] T. Ikeda and M. Nagahara, "Value function in maximum hands-off control for linear systems," *Automatica*, vol. 64, pp. 190–195, 2016.

[56] ——, "Discrete-valued model predictive control using sum-of-absolute-values optimization," *Asian Journal of Control*, vol. 20, no. 1, pp. 196–206, 2018.

[57] ——, "Time-optimal hands-off control for linear time-invariant systems," *Automatica*, vol. 99, pp. 54–58, 2019.

[58] M. Ishikawa, "Structural learning with forgetting," *Neural Netw.*, vol. 9, no. 3, pp. 509–521, Apr. 1996.

[59] A. Jadbabaie, A. Olshevsky, G. J. Pappas, and V. Tzoumas, "Minimal reachability is hard to approximate," *IEEE Trans. Autom. Control*, vol. 64, no. 2, pp. 783–789, 2019.

[60] D. Jeong and W. Jeon, "Performance of adaptive sleep period control for wireless communications systems," *IEEE Trans. Wireless Commun.*, vol. 5, no. 11, pp. 3012–3016, Nov. 2006.

[61] M. R. Jovanović and N. K. Dhingra, "Controller architectures: Tradeoffs between performance and structure," *European Journal of Control*, vol. 30, pp. 76 – 91, 2016, 15th European Control Conference, ECC16.

[62] N. Karumanchi, *Data Structures and Algorithms Made Easy*, 2nd ed. Career-Monk, 2011.

[63] E. Khmelnitsky, "On an optimal control problem of train operation," *IEEE Trans. Autom. Control*, vol. 45, no. 7, pp. 1257–1266, 2000.

[64] R. Kirchhoff, M. Thele, M. Finkbohner, P. Rigley, and W. Settgast, "Start-stop system distributed in-car intelligence," *ATZextra worldwide*, vol. 15, no. 11, pp. 52–55, Jan. 2010.

[65] E. Kreyszig, *Introductory Functional Analysis with Applications*. Wiley, 1989.

[66] D. Liberzon, *Calculus of Variations and Optimal Control Theory: A Concise Introduction*. Princeton University Press, 2012.

[67] F. Lin, M. Fardad, and M. R. Jovanović, "Augmented Lagrangian approach to design of structured optimal state feedback gains," *IEEE Trans. Autom. Control*, vol. 56, no. 12, pp. 2923–2929, 2011.

[68] F. Lin, M. Fardad, and M. R. Jovanović, "Design of optimal sparse feedback gains via the alternating direction method of multipliers," *IEEE Trans. Autom. Control*, vol. 58, no. 9, pp. 2426–2431, 2013.

[69] B. F. Logan, "Properties of high-pass signals," Ph.D. dissertation, Columbia University, 1965.

[70] J. M. Maciejowski, *Predictive Control with Constraints*. Prentice-Hall, 2002.

[71] S. G. Mallat and Z. Zhang, "Matching pursuits with time-frequency dictionaries," *IEEE Trans. Signal Process.*, vol. 41, no. 12, pp. 3397–3415, Nov. 1993.

[72] S. Mallat, *A Wavelet Tour of Signal Processing: The Sparse Way*, 3rd ed. Academic Press, 2008.

[73] I. Markovsky, *Low Rank Approximation*. Springer, 2012.

[74] R. Martin, K. L. Teo, and M. D'Incalci, *Optimal Control of Drug Administration in Cancer Chemotherapy*. Singapore: World Scientific, 1994.

[75] M. Mesbahi and G. P. Papavassilopoulos, "On the rank minimization problem over a positive semidefinite linear matrix inequality," *IEEE Trans. Autom. Control*, vol. 42, no. 2, pp. 239–243, Feb. 1997.

[76] U. Münz, M. Pfister, and P. Wolfrum, "Sensor and actuator placement for linear systems based on H_2 and H_∞ optimization," *IEEE Trans. Autom. Control*, vol. 59, no. 11, pp. 2984–2989, 2014.

[77] M. Nagahara, "Discrete signal reconstruction by sum of absolute values," *IEEE Signal Process. Lett.*, vol. 22, no. 10, pp. 1575–1579, Oct. 2015.

[78] M. Nagahara and C. F. Martin, "Monotone smoothing splines using general linear systems," *Asian Journal of Control*, vol. 5, no. 2, pp. 461–468, Mar. 2013.

[79] ——, "L^1 control theoretic smoothing splines," *IEEE Signal Process. Lett.*, vol. 21, no. 11, pp. 1394–1397, Nov. 2014.

[80] M. Nagahara, T. Matsuda, and K. Hayashi, "Compressive sampling for remote control systems," *IEICE Trans. on Fundamentals*, vol. E95-A, no. 4, pp. 713–722, Apr. 2012.

[81] M. Nagahara and D. E. Quevedo, "Sparse representations for packetized predictive networked control," in *IFAC 18th World Congress*, Aug.–Sept. 2011, pp. 84–89.

[82] M. Nagahara, D. E. Quevedo, and D. Nešić, "Maximum hands-off control and L^1 optimality," in *52nd IEEE Conference on Decision and Control (CDC)*, Dec. 2013, pp. 3825–3830.

[83] ——, "Maximum hands-off control: a paradigm of control effort minimization," *IEEE Trans. Autom. Control*, vol. 61, no. 3, pp. 735–747, 2016.

[84] M. Nagahara, D. Quevedo, and J. Østergaard, "Sparse packetized predictive control for networked control over erasure channels," *IEEE Trans. Autom. Control*, vol. 59, no. 7, pp. 1899–1905, Jul. 2014.

[85] M. Nagahara, D. Chatterjee, N. Challapalli, and M. Vidyasagar, "CLOT norm minimization for continuous hands-off control," *Automatica*, vol. 113, p. 108679, 2020.

[86] M. Nagahara, J. Østergaard, and D. E. Quevedo, "Discrete-time hands-off control by sparse optimization," *EURASIP Journal on Advances in Signal Processing*, vol. 2016, no. 1, pp. 1–8, 2016.

[87] M. Nalbach, A. Korner, and S. Kahnt, "Active engine-off coasting using 48V: Economic reduction of CO_2 emissions," in *17th International Congress ELIV*, Oct. 2015, pp. 41–51.

[88] D. Needell and J. A. Tropp, "CoSaMP: iterative signal recovery from incomplete and inaccurate samples," *Appl. Comput. Harmonic Anal.*, vol. 26, no. 3, pp. 301–321, 2008.

[89] K. Ogata, *Modern Control Engineering*, 5th ed. Pearson, 2009.

[90] A. Olshevsky, "Minimal controllability problems," *IEEE Trans. Control Netw. Syst.*, vol. 1, no. 3, pp. 249–258, 2014.

[91] S. K. Pakazad, H. Ohlsson, and L. Ljung, "Sparse control using sum-of-norms regularized model predictive control," in *52nd IEEE Conference on Decision and Control*, 2013, pp. 5758–5763.

[92] N. Parikh and S. Boyd, "Proximal algorithms," *Foundations and Trends in Optimization*, vol. 1, no. 3, pp. 123–231, 2013.

[93] F. Pasqualetti, S. Zampieri, and F. Bullo, "Controllability metrics, limitations and algorithms for complex networks," *IEEE Trans. Control Netw. Syst.*, vol. 1, no. 1, pp. 40–52, 2014.

[94] Y. C. Pati, R. Rezaiifar, and P. S. Krishnaprasad, "Orthogonal matching pursuit: Recursive function approximation with applications to wavelet decomposition," in *Proc. the 27th Annual Asilomar Conf. on Signals, Systems and Computers*, Nov. 1993, pp. 40–44.

[95] S. Pequito, S. Kar, and A. P. Aguiar, "A framework for structural input/output and control configuration selection of large-scale systems," *IEEE Trans. Autom. Control*, vol. 61, no. 2, pp. 303–318, Feb. 2016.

[96] L. S. Pontryagin, *Mathematical Theory of Optimal Processes*. CRC Press, 1987, vol. 4.

[97] L. I. Rudin, S. Osher, and E. Fatemi, "Nonlinear total variation based noise removal algorithms," *Physica D*, vol. 60, p. 259–268, 1992.

[98] W. Rudin, *Principles of Mathematical Analysis*, 3rd ed. McGraw-Hill, 1976.

[99] ——, *Real and Complex Analysis*, 3rd ed. McGraw-Hill, 2005.

[100] F. Santosa and W. W. Symes, "Linear inversion of band-limited reflection seismograms," *SIAM Journal on Scientific and Statistical Computing*, vol. 7, no. 4, pp. 1307–1330, 1986.

[101] H. Sasahara, K. Hayashi, and M. Nagahara, "Symbol detection for faster-than-Nyquist signaling by sum-of-absolute-values optimization," *IEEE Signal Process. Lett.*, vol. 23, no. 12, pp. 1853–1857, 2016.

[102] ——, "Multiuser detection based on MAP estimation with sum-of-absolute-values relaxation," *IEEE Trans. Signal Process.*, vol. 65, no. 21, pp. 5621–5634, 2017.

[103] H. Schättler and U. Ledzewicz, *Geometric Optimal Control*. Springer, 2012.

[104] B. Schölkopf and A. J. Smola, *Learning with Kernels*. The MIT Press, 2002.

[105] P. Shakouri, A. Ordys, P. Darnell, and P. Kavanagh, "Fuel efficiency by coasting in the vehicle," *International Journal of Vehicular Technology*, vol. 2013, p. 14, 2013.

[106] N. Srivastava, G. Hinton, A. Krizhevsky, I. Sutskever, and R. Salakhutdinov, "Dropout: A simple way to prevent neural networks from overfitting," *Journal of Machine Learning Research*, vol. 15, pp. 1929–1958, 2014.

[107] G. Strang and T. Nguyen, *Wavelets and Filter Banks*, 2nd ed. Wellesley-Cambridge Press, 1996.

[108] S. Sun, M. B. Egerstedt, and C. F. Martin, "Control theoretic smoothing splines," *IEEE Trans. Autom. Control*, vol. 45, no. 12, pp. 2271–2279, Dec. 2000.

[109] H. L. Taylor, S. C. Banks, and J. F. McCoy, "Deconvolution with the ℓ_1 norm," *Geophysics*, vol. 44, no. 1, pp. 39–52, 1979.

[110] R. Tibshirani, "Regression shrinkage and selection via the LASSO," *J. R. Statist. Soc. Ser. B*, vol. 58, no. 1, pp. 267–288, 1996.

[111] V. Tzoumas, M. A. Rahimian, G. J. Pappas, and A. Jadbabaie, "Minimal actuator placement with bounds on control effort," *IEEE Trans. Control Netw. Syst.*, vol. 3, no. 1, pp. 67–78, 2016.

[112] M. Vidyasagar, *An Introduction to Compressed Sensing*. SIAM, 2019.

[113] G. Vossen and H. Maurer, "On L^1-minimization in optimal control and applications to robotics," *Optimal Control Applications and Methods*, vol. 27, no. 6, pp. 301–321, 2006.

[114] Y. Yamamoto, *From Vector Spaces to Function Spaces: Introduction to Functional Analysis with Applications*. SIAM, 2012.

[115] N. Young, *An Introduction to Hilbert Space*. Cambridge University Press, 1988.

[116] M. Yuan and Y. Lin, "Model selection and estimation in regression with grouped variables," *Journal of the Royal Statistical Society: Series B (Statistical Methodology)*, vol. 68, no. 1, p. 49–67, Feb. 2006.

[117] K. Zhou, J. C. Doyle, and K. Glover, *Robust and Optimal Control*. Pearson, 1995.

[118] M. Zibulevsky and M. Elad, "L1-L2 optimization in signal and image processing," *IEEE Signal Process. Mag.*, vol. 27, pp. 76–88, May 2010.

[119] H. Zou and T. Hastie, "Regularization and variable selection via the elastic net," *Journal of the Royal Statistical Society: Series B (Statistical Methodology)*, vol. 67, no. 2, pp. 301–320, Apr. 2005.

Index